TIRES AND TRACKS

A basic guide to the inspection, repair, and maintenance of tires and tracks for off-road vehicles

FUNDAMENTALS OF SERVICE

PUBLISHER
DEERE & COMPANY
JOHN DEERE PUBLISHING
one John Deere Place
Moline, IL 61265
http://www.johndeere.com/publications
1–800–522–7448

Fundamentals of Service (FOS) is a series of manuals created by Deere & Company. Each book in the series is conceived, researched, outlined, edited, and published by Deere & Company, John Deere Publishing. Authors are selected to provide a basic technical manuscript that could be edited and rewritten by staff editors.

HOW TO USE THE MANUAL: This manual can be used by anyone — experienced mechanics, shop trainees, vocational students, and lay readers.

Persons not familiar with the topics discussed in this book should begin with Chapter 1 and then study the chapters in sequence. The experienced person can find what is needed on the "Contents" page.

Each guide was written by Deere & Company, John Deere Publishing staff in cooperation with the technical writers, illustrators, and editors at Almon, Inc. — a full-service technical publications company headquartered in Waukesha, Wisconsin (www.almoninc.com).

FOR MORE INFORMATION: This book is one of many books published on agricultural and related subjects. For more information or to request a FREE CATALOG, call 1–800–522–7448 or send your request to address above or:

Visit Us on the Internet
http://www.johndeere.com/publications

We have a long-range interest in Vocational Education

ISBN 0-86691-386-6

CONTENTS

1 PNEUMATIC TIRES

2 METAL TRACKS

3 RUBBER TRACKS

4 APPENDIX

PNEUMATIC TIRES

Tires

This manual is divided into two parts. Part 1 covers construction, identification, operation, service and maintenance of pneumatic tires. Part 1 consists of Chapter 1, "Pneumatic Tires".

Part 2 consists of Chapters 2 and 3 and will cover traditional and rubber tracks.

Pneumatic Rubber Tires

Pneumatic rubber tires are used on almost all wheel machines.

These tires are designed for two general uses:

- Over-The-Road — for speeds over 30 mph (48 km)
- Off-The-Road — for speeds less than 30 mph (48 km)

Over-the-road tires flex faster and generate more heat than off-the-road tires.

Off-the-road tires operate at lower speeds and generate less heat. But they must be tougher to take shocks from rocks, stumps, and bumps.

Some tires, such as truck tires, are designed to satisfy both conditions.

Within each category, a wide variety of tires is available: different sizes, plies, profiles, and treads.

This chapter is not intended to be the final word on tires. Always follow tire, wheel, and rim manufacturer's and manufacturer's association handling and repair procedures.

Construction of Tires

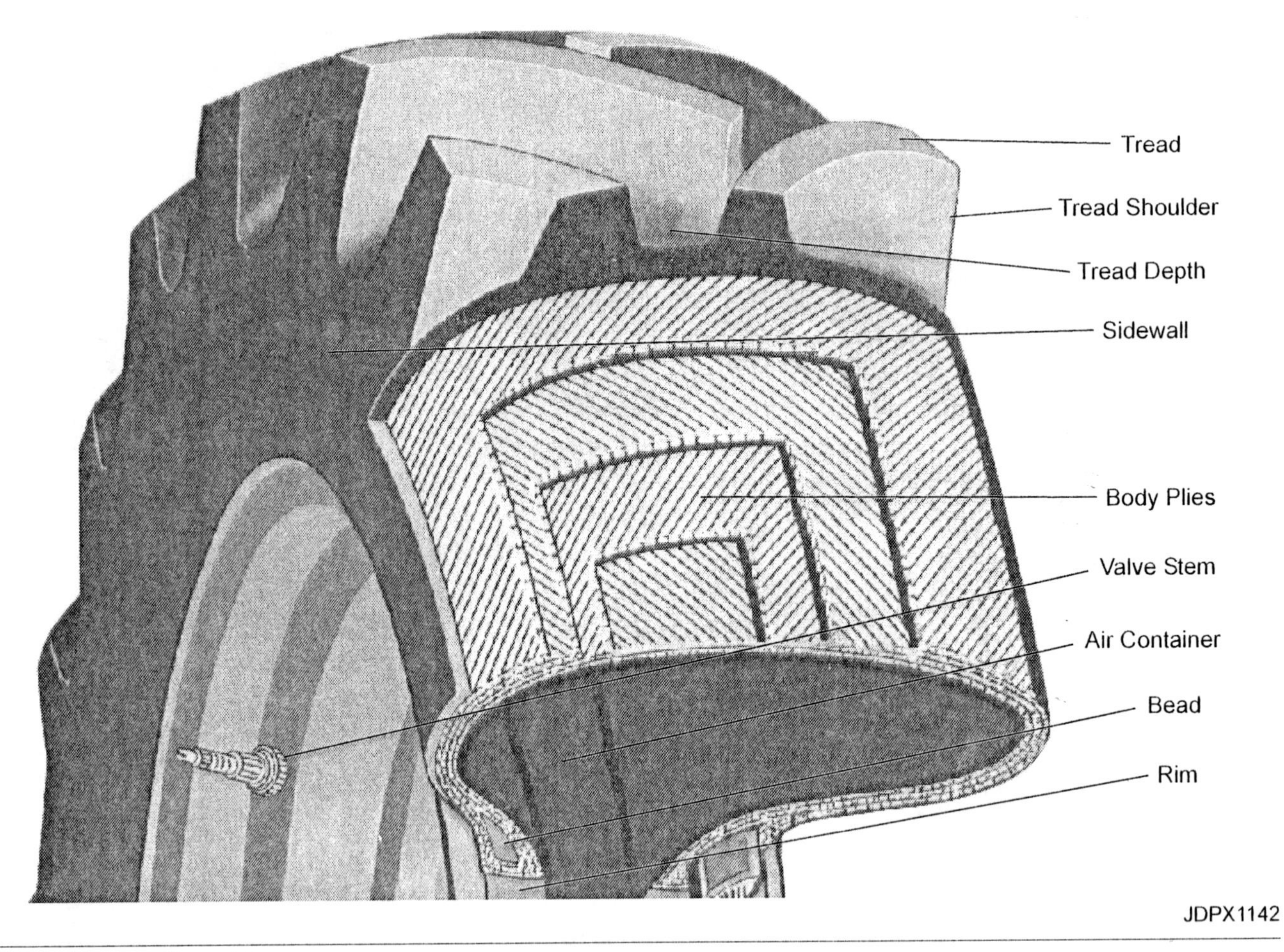

Fig. 1 — Cutaway View of Tire

You can see the vital parts of a pneumatic tire in Fig. 1. The function of each part is given below.

BEAD

Tire beads anchor the tire to the rim. They are bundles of wire that are the foundation of the tire. All the tire plies are tied to the beads and the wire beads prevent changes of shape or fit on the rim.

BODY PLIES

Body plies are layers of rubber-cushioned cord or fabric. The body must be strong enough to hold in the inflation pressure which supports the load and cushions the shocks. Each cord in each ply is surrounded by a resilient rubber compound and each ply is insulated from the next by a layer of the same compound.

The cord material may be cotton, rayon, nylon, polyester, steel, etc. Nylon is the most popular cord material; however, steel cord is used in many radial tires.

A tire may have two, four, or six plies of cord (usually for automobiles, sport utility vehicles, and light pick-up trucks); or six to 14 for large trucks. Sometimes 20 or more plies are built into tires for large off-the-road equipment.

PLY RATING

The carcass strength of tires for off-the-road use is indicated by a "ply rating" for bias and belted bias ply tires. "Star markings" are used on radial tires to indicate the tire's strength.

At one time, the ply rating used on bias ply and belted bias ply tires told the actual number of plies in the body of the tire. Now it identifies a given tire with its recommended load when used in a specific type of service.

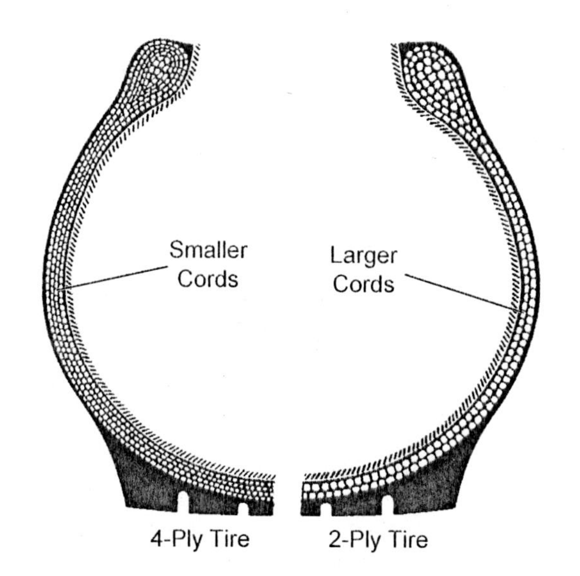

Fig. 2 — Construction of 2- and 4-Ply Tires Compared

Ply rating tells the strength, but does not necessarily indicate the actual number of cord plies in the tire. For example, many tires with a 4-ply rating may only contain two plies. This simply means that the plies have added strength and will carry the load of a tire that has four plies of smaller diameter cords.

A comparison between 2-ply and 4-ply tire construction is shown in Fig. 2. To get the same strength, note how the 2-ply tire cords are larger.

Fig. 3 — Radial Tire Identification Panel

NOTE: Automotive and truck tires are now identified by "load range" rather than ply rated.

Radial tires are marked with either one, two, or three star symbols (★) to designate the load capacity of the tire. A panel cured in the sidewall of the tire will show the symbol marking (Fig. 3).

Maximum load rating for all radial tires with one star (★) is at 18 psi (124 kPa) inflation pressure. Maximum load rating for all tires with two stars (★★) is at 24 psi (165 kPa) and for all tires with three stars (★★★) is at 30 psi (207 kPa). The tire will also be marked to indicate the maximum load the tire will carry at the appropriate inflation pressure, and which bias ply rating the radial tire will replace.

Bias Ply

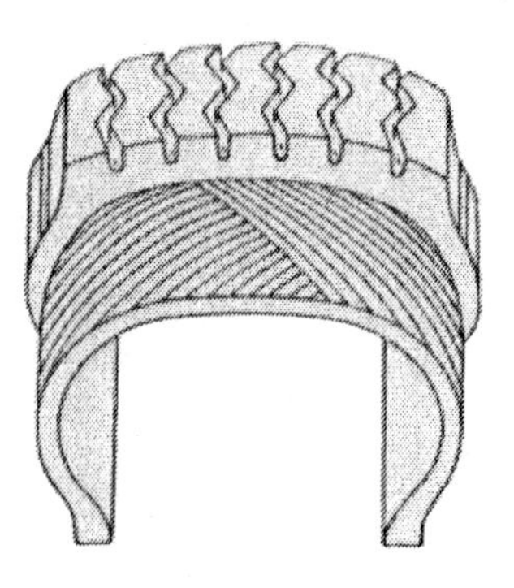

Fig. 4 — Bias Ply Construction of Tires

In normal bias ply tire construction, the ply cords run from one tire bead to the other at an angle (Fig. 4).

Alternate plies of the body run in opposite directions. This bias construction gives rigidity to both the sidewall and the tread.

Belted Bias Ply

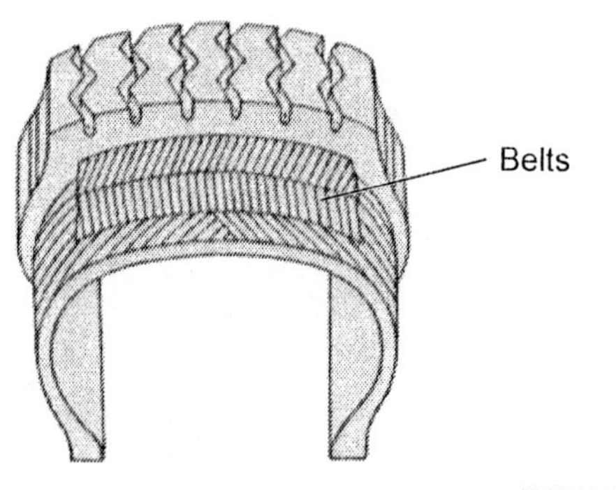

Fig. 5 — Belted Bias Ply Construction of Tires

In belted bias, the tire body is similar to that of the bias tire, except the body is surrounded by fairly rigid belts (Fig. 5). These belts are composed of cords which surround the tire body underneath the tread. The cords have much less angle than the cords in bias ply tires. This arrangement gives rigidity to the sidewalls and even more rigidity to the tread. The belts reduce tread motion during contact with the road, thus improving tread life.

Belted Radial Ply

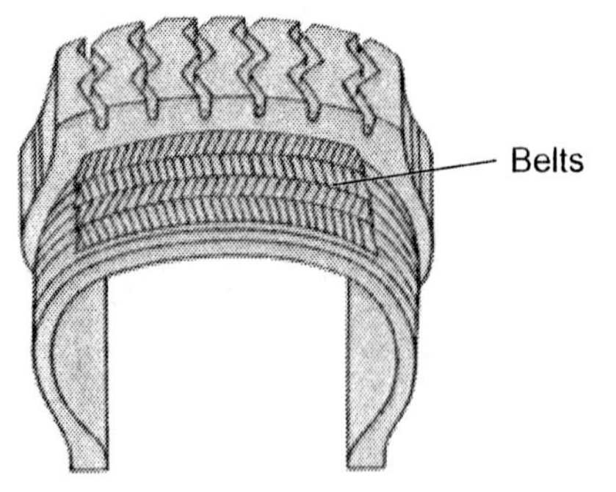

JDPX1404

Fig. 6 — Belted Radial Ply Construction of Tires

Here the body cords run across the body from bead to bead at almost a right angle (Fig. 6). Fairly rigid belts, composed of cords, surround the tire body underneath the tread. As in belted bias tires, the belts reduce tread motion during contact with the road, thus improving life. Radial construction gives better support to the tread than bias ply or belted bias ply tires.

Wire-Reinforced Ply

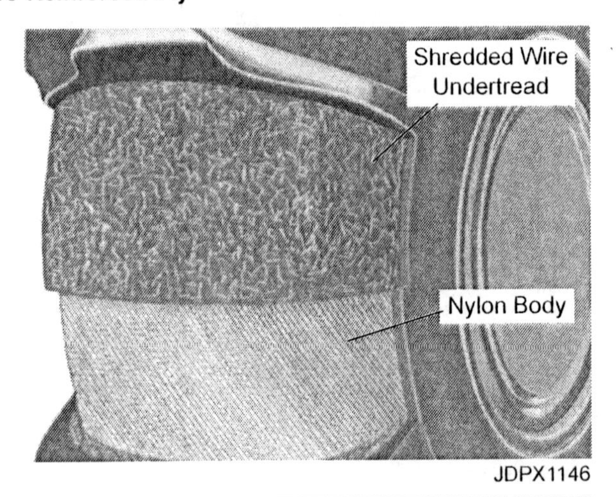

JDPX1146

Fig. 7 — Wire-Reinforced Construction of Heavy-Duty Tire

In some big tires, for earthmoving and other heavy equipment, a layer of wire is cured between the tread and the body of the tire (Fig. 7). This protective layer of wire keeps most cuts from penetrating into the body, keeps tread cuts from growing, and holds tread cuts closed so sand and dirt don't enter and cause separation.

Some tires are marked "SWB" (shredded wire braid) on the sidewall.

SIDEWALLS

Sidewalls are the rubber coverings on both sides of the body. They are designed to flex and bend without cracking during both ordinary deflection and extreme sudden shock.

TREAD

The tread is the part of the tire which contacts the road. It must provide traction, provide long wear, and resist cuts. There are many tread patterns and depths.

TUBELESS INNER LINER

The liner (not shown) is an integral part of the tubeless tire, covering the inside from bead to bead. The liner in the tire serves as the air retainer. It reduces weight by eliminating tubes and flaps and simplifies maintenance.

TUBE

The function of the inner tube is to hold air, inert gases, or liquid under pressure in a tube-type tire.

FLAPS

Tires with tubes have flaps to protect the tube from injury by contact with the rim and tire beads.

RIM

The rim supports the tire.

Codes for Tires

JDPX1147

Fig. 8 — Code for Tire Type Stamped on Side of Tire

Many kinds of tires are made, especially for agriculture. For simplicity, a uniform code has been adopted by both the Tire and Rim Association and The Rubber Manufacturers Association (RMA).

The standard code is marked on the side of the tire and normally consists of a letter plus a number (Fig. 8). The letter "R" denotes a rear tractor tire... the number "1" following tells you the tire has regular tread.

See the following tire code chart. The RMA codes provide a guide for determining the type of service for which a tire is intended. Using a tire for any service that it is not designed for will result in poor performance and possibly short tire life.

STANDARD INDUSTRY CODES FOR TIRE TYPES*	
Type of Tire	**Code**
FRONT TRACTOR	
Rice tread	F-1
Single rib tread	F-2
Dual rib tread	F-2D
Triple rib tread	F-2T
Industrial tread	F-3
DRIVE WHEEL TRACTOR (REAR)	
Rear wheel, regular tread	R-1
Cane and Rice, deep tread	R-2 **
Shallow, non-directional tread	R-3
Industrial, intermediate tread	R-4
IMPLEMENT	
Rib tread	I-1
Traction tread	I-3
Plow tailwheel	I-4
Smooth tread	I-6
OFF-THE-ROAD TIRES (INDUSTRIAL)	
Rib	E-1
Traction	E-2
Rock	E-3
Rock deep tread	E-4
Rock intermediate	E-5
Rock maximum	E-6
Flotation	E-7
Also includes similar treads for "G," "L," and "ML" series codes.	

* *Few tire manufacturers make every kind of tire. If there is no code, a tire is marked with a code for the most similar tire. If you are not certain which tire is best for your purpose, consult a reputable dealer for advice.*

** *Some manufacturers add the letter "O" after the industry code to denote an "open tread" design, and the letter "C" to denote a "closed tread" design in the R-2 classification.*

Tire Sizes — What They Mean

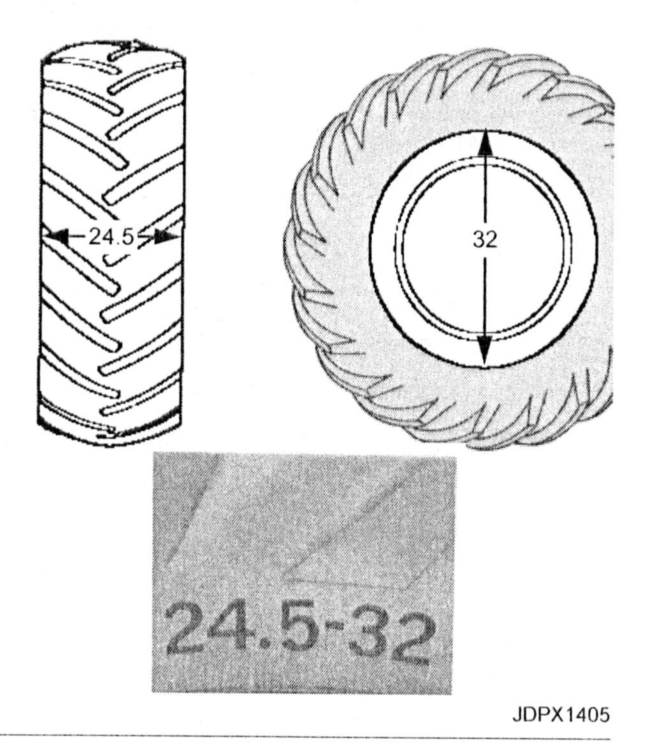

Fig. 9 — Tire Sizes Marked on Sidewall of Tires

A typical size marking is located on the sidewall of an off-the-road tire (Fig. 9). The "24.5" tells the section width of the tire in inches when mounted on its recommended rim. The "32" gives the nominal diameter (in inches) of the rim.

The tire size designation of most radial tires carries metric tire width measurements. An example of the metric tire size marking is the designation "420/80R46." The "420" tells the section width of the tire in millimeters (approximately 16.5 inches). The "80" designation is the aspect ratio of the tire. The aspect ratio designation is determined by dividing the total height of the tire cross section by its width. The lower the aspect ratio, the shorter and wider the tire will be. While a higher aspect ratio designates a taller and narrower tire. For example, a "70" aspect ratio tire would stand shorter and wider than an "80" tire. The "R" indicates radial-ply construction. The "46" designates the rim diameter in inches.

CAUTION: Never try to fit a tire to a rim that does not exactly match the specified rim diameter molded on the tire sidewall.

DUAL SIZING FOR TRACTOR DRIVE WHEEL TIRES

Several years ago, extra wide base rims were adopted for rear tractor tires. They permitted the tire beads to be spaced wider apart on the rim giving the tire a wider, more stable base and a wider cross section.

Prior to wide rims, tire size markings were general and did not give the tire section width. An 11-28 tire actually had a tire section width of 11.90 inches when mounted on a 10-inch-wide rim.

An example of dual sizing is the designation "12.4/11-28." The section width is given with the "12.4" designation. The "11" designation is the old sizing.

Gradually, as the older, narrower rims become obsolete, dual size will be eliminated and only the size for wide base rims will be shown (as pictured for the 24.5-32 tire size).

Directional Arrows on Tires

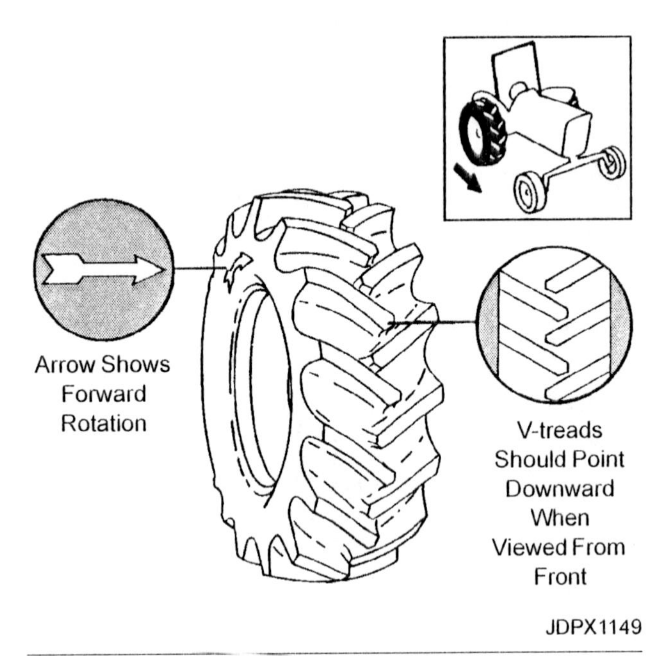

Fig. 10 — Arrow on Tire Indicates Forward Rotation

Drive wheel tires on many machines must be mounted to rotate in a certain direction for best traction. For this reason, arrows are often stamped on the sidewall to show forward rotation (Fig. 10). This is most important on agricultural V-tread tires where the "V" pattern must point downward when viewed from the front.

Reversing the tread will damage the tread bars. Only on ground-driven implements such as planters and spreaders should the V-tread be reversed.

Tire Inflation

The most important part of tire maintenance is maintaining the proper inflation pressure. Tires are designed to operate with a certain sidewall deflection or "bulge." Correct air pressure is necessary for this bulge, and for proper traction, flotation, load support and flex control. Knowing the weight the tire must support, you can adjust the tire inflation pressure to achieve a high level of traction by maintaining as large a contact area, or "footprint," between the tire and soil as possible. This is called operating at the "rated deflection" of the tire.

NOTE: When a tire inflation pressure is discussed here, it will be given in pounds per square inch (psi) first, followed by the metric equivalent, which is given is kilopascals (kPa) and shown in parenthesis. For example: inflate to 24 psi (165 kPa).

Radial Tires

Radial tires provide improved traction and fuel economy, reduced soil compaction, improved flotation, a smoother ride and longer tire life compared to bias ply tires as a result of their high contact area between the rubber and the soil. Proper adjustment of radial tires to achieve this high contact area is obtained by using the correct tire inflation pressure to support the static load being carried by the tire for a given field condition.

When setting up a tractor for heavy tillage operation, the bigger the tire air volume the better. The tires will provide the best performance, and hop can be readily controlled when the tractor is equipped with tires that are large enough relative to the weight of the tractor to provide a "soft" ride (low sidewall stiffness). The tires should be inflated to the minimum pressure required to support the static weight being carried on each axle.

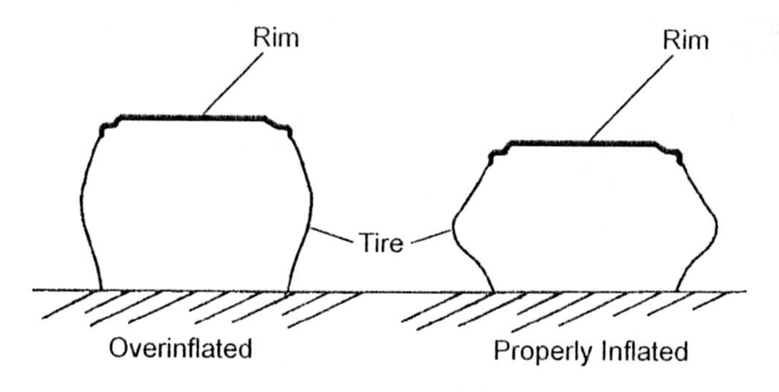

Fig. 11 — A Properly Inflated Radial Tire Will Have More Sidewall Deflection Than Comparable Bias Ply Tire

Depending on the weight of the tractor, as little as 6 psi (41 kPa) inflation pressure may be needed to provide optimum tire performance. This will result in a tire that might look underinflated or "flat," but the bulge in a properly inflated radial tire in normal and is needed to maintain the optimum "footprint" (Fig. 11).

It may even look like the tire will slip on the rim under load at these pressures, but this is not a significant problem. Modern tires and rims are manufactured for a closer fit in the bead seat area, and with the larger rim diameters that are commonly used, rim slip is rare under normal field conditions.

The essential requirement for preventing rim slip is that proper tire mounting procedures be followed, using only a minimum amount of the specified tire lubricant to seat the tire beads.

Conventional Size Radial Drive Wheel Tires

Tire Size Designation	Inflation (psi)	6	7	8	9	10	12	14	16	18	20	22	24	26	28	30
18.4R38	Single (lbs.)	NR	NR	3520	3740	3960	4400	4800	5200	5680	6000	6400	6600			
	Dual (lbs.)	2640	2820	3100	3290	3480	3870	4220	4580	5000	5280	5630	5810			
	Triple (lbs.)	2460	2620	2890	3070	3250	3610	3940	4260	4660	4920	5250	5410			
18.4R42	Single (lbs.)	NR	NR	3740	3960	4180	4680	5080	5520	6000	6400	6600	6950			
	Dual (lbs.)	2710	3010	3290	3480	3680	4120	4470	4860	5280	5630	5810	6120			
	Triple (lbs.)	2530	2800	3070	3250	3430	3840	4170	4530	4920	5250	5410	5700			
18.4R46	Single (lbs.)	NR	NR	3860	4180	4400	4940	5360	5840	6150	6600	6950	7400	7850	8050	8550
	Dual (lbs.)	2900	3200	3400	3680	3870	4350	4720	5140	5410	5810	6120	6510	6910	7080	7520
	Triple (lbs.)	2710	2980	3170	3430	3610	4050	4400	4790	5040	5410	5700	6070	6440	6600	7020
20.8R38	Single (lbs.)	NR	NR	4300	4540	4800	5360	5840	6400	6800	7150	7600	8050			
	Dual (lbs.)	3200	3480	3780	4000	4220	4720	5140	5630	5980	6290	6690	7080			
	Triple (lbs.)	2980	3250	3530	3720	3940	4400	4790	5250	5580	5860	6230	6600			
20.8R42	Single (lbs.)	NR	NR	4540	4800	5080	5680	6150	6800	7150	7600	8050	8550			
	Dual (lbs.)	3290	3680	4000	4220	4470	5000	5410	5980	6290	6690	7080	7520			
	Triple (lbs.)	3070	3430	3720	3940	4170	4660	5040	5580	5860	6230	6600	7010			
24.5R32	Single (lbs.)	NR	NR	5080	5520	5840	6400	7150	7600	8250	8800	9100	9650			
	Dual (lbs.)	3780	4120	4470	4860	5140	5630	6290	6690	7260	7740	8010	8490			
	Triple (lbs.)	3530	3840	4170	4530	4790	5250	5860	6230	6770	7220	7460	7910			
30.5LR32	Single (lbs.)	NR	NR	6150	6600	6950	7600	8550	9100	9650						
	Dual (lbs.)	4470	5000	5410	5810	6120	6690	7520	8010	8490						
	Triple (lbs.)	4170	4660	5040	5410	5700	6230	7010	7460	7910						

NR – Not Recommended
Goodyear Tire & Rubber Company

Metric Size Radial Drive Wheel Tires

Tire Size Designation	Inflation (psi)	6	7	9	10	12	13	15	17	20	23	26	29
420/80R46	Single (lbs.)	NR	NR	3520	3860	4080	4400	4680	5360	5840	6400	6800	7150
	Dual (lbs.)	2560	2820	3100	3400	3590	3870	4120	4720	5140	5630	5980	6290
	Triple (lbs.)	2390	2620	2890	3170	3350	3610	3840	4400	4790	5250	5580	5860
710/70R38	Single (lbs.)	NR	NR	2900	3150	3350	3650	3875	4375	4875	5300		
	Dual (lbs.)	2140	2330	2550	2770	2950	3210	3410	3850	4290	4665		
	Triple (lbs.)	1995	2175	2380	2585	2745	2995	3180	3590	4000	4345		

NR – Not Recommended
Goodyear Tire & Rubber Company

JDPX1151

Fig. 12 — Extended Inflation Tables for Radial Drive Wheel Tires

Previously, tire inflation tables did not extend below 12 psi (82 kPa), and many operator's manuals warned not to go below inflation pressures of 12 psi (82 kPa). The load and inflation tables for large radial tires shown in Fig. 12 have been extended downward to accommodate the lower pressures now recommended by tire and equipment manufacturers.

To use the tables, first determine the static weight on each axle. Then, divide the weight per axle by the number of tires on the axle to determine the weight being supported by the tire. Find the nearest weight for your tire size and read the recommended inflation pressure at the top of the tables. Note that the sidewall deflection for each tire size will be about the same going across a given row in the table — you will get the same amount of "bulge" under the specified weight load using the recommended minimum inflation pressure. This is the proper inflation pressure for radial tires to operate at the rated deflection for optimum performance.

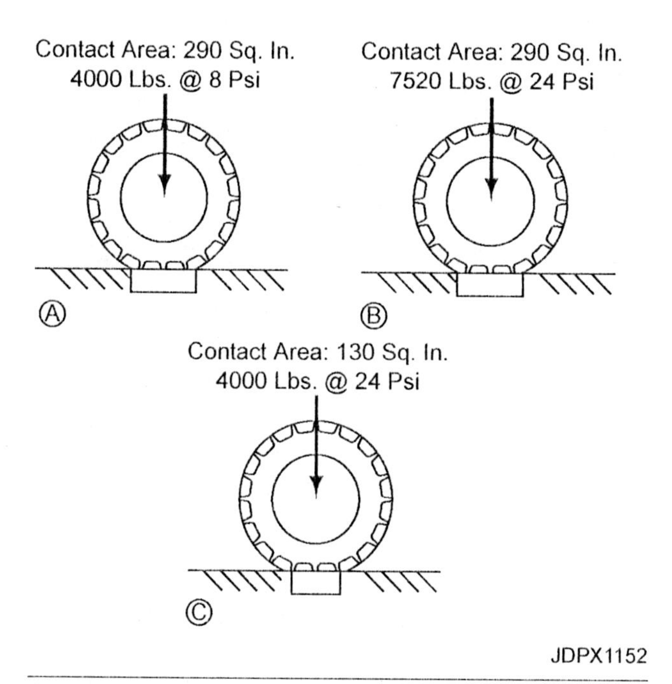

Fig. 13 — How Air Pressure and Load Affects Tire Footprint Size

As an example, a 20.8R42 rear tractor tire mounted as a dual will support 4,000 pounds of static load when inflated to 8 psi (56 kPa) (Fig. 13, A), but at 24 psi (165 kPa) it will support 7,520 pounds (B). Under these two different inflation and load levels, the tire has the same size footprint — 290 square inches. This size footprint is the rated deflection for a 20.8R42 tire and will provide optimum tire performance. However, inflate the tire to 24 psi (165 kPa) under the 4,000 pound load and the footprint shrinks to 130 square inches (C). The smaller footprint will result in a substantial reduction in tire performance, and the potential for wheel hop and soil compaction would be greater.

It is important to note that when tractor ballasting and tire inflation are optimized for one service category, such as towing high-draft implements, switching to another category, such as hitch-mounted implements, may require ballast changes and inflation pressure changes to maintain optimum tire performance.

For mechanical front wheel drive radial front tires, two-wheel drive row crop front tires and any bias ply tires, continue to use the appropriate inflation guidelines developed by the American Society of Agricultural Engineers to determine the proper inflation pressure. See your tire dealer.

Bias Ply Tires

Bias ply tires have stiffer sidewalls than radial tires, and therefore do not have a distinct "bulge" in the sidewall when inflated to the proper pressure for the load being carried as do radial tires. However, maintaining the correct tire inflation pressure is equally important for bias ply tires as it is for radial tires to obtain optimum performance and long tire life.

OVERINFLATION

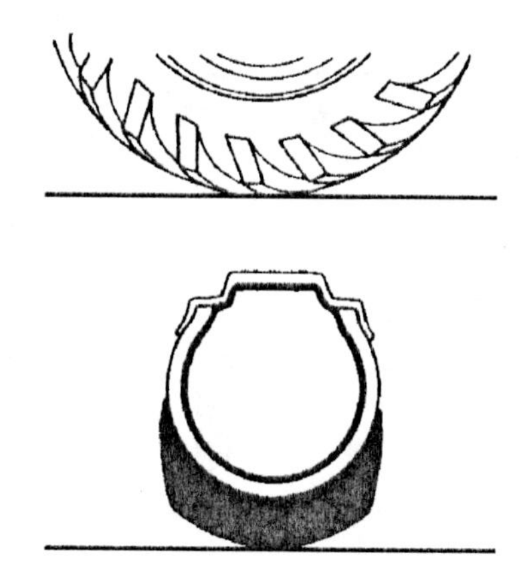

Fig. 14 — Overinflated Tire

Overinflated tires don't fully contact the ground (Fig. 14). The center of the tread wears. And, because the tire is more rigid, it is more liable to be damaged by curbs, rocks, etc.

CAUTION: Overinflation and overloading my lead to a violent tire explosion which may result in severe equipment damage, serious injury and death.

UNDERINFLATION

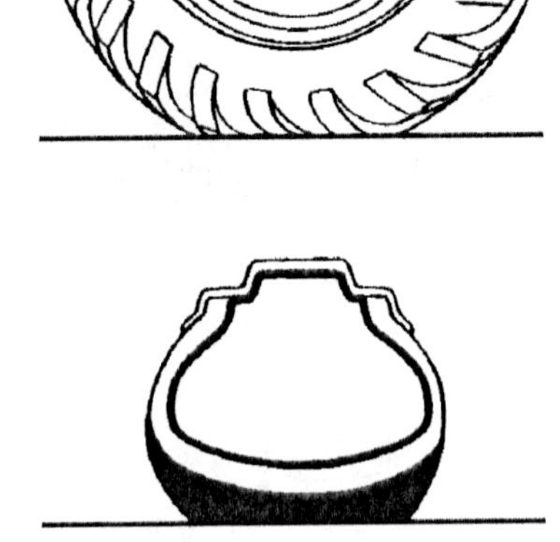

JDPX1407

Fig. 15 — Underinflated Tire

Underinflated tires flex excessively (Fig. 15) every time the wheel turns, resulting in high internal heat and premature failure. Underinflation is indicated by excessive wear of the tread on the sides while the center is relatively unworn.

PROPER INFLATION

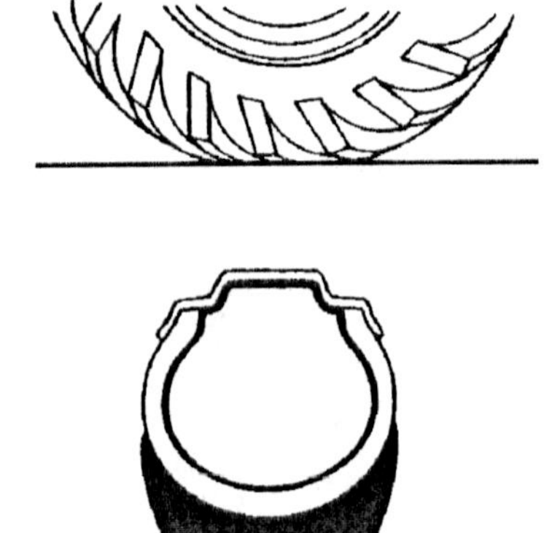

JDPX1408

Fig. 16 — Correctly Inflated Tire

Correctly inflated tires (Fig. 16) permit all of the tread to contact the ground, yet are not soft enough to flex excessively.

NITROGEN INFLATION

Some original equipment manufacturers of off-the-road equipment recommend nitrogen inflation of tires. The nitrogen helps to minimize the possibility of an explosion due to excessive heat from external sources. Nitrogen inflation also reduces the aging of the tire due to oxidation of the tire carcass. It also minimizes rust.

PRESSURE GAUGE

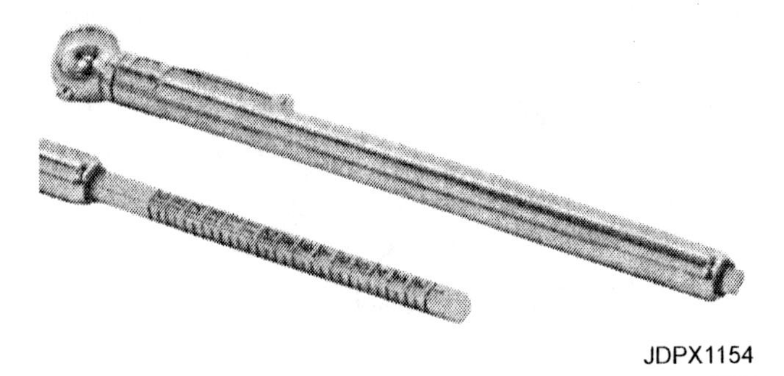

JDPX1154

Fig. 17 — Low-Pressure Tire Gauge

Proper inflation must be maintained at all times and all seasons. An accurate pressure gauge is required. Two types of pressure gauges are available for either dry or liquid testing: high-pressure and low-pressure. Always use the proper gauge when checking pressure. Normally, high-pressure gauges are not satisfactory for checking low-pressure tires, as the gauges are usually calibrated in five-pound increments. Low-pressure gauges are better for low pressure; they are calibrated in one-pound increments.

Accuracy is most important. Accuracy can only be determined by checking a gauge against a gauge of known accuracy. For this reason, keep one accurate gauge for the sole purpose of checking the accuracy of other gauges. Low-pressure gauges, Fig. 17, are especially susceptible to inaccuracy.

HOW TO INFLATE TIRES

CAUTION: Always wear protective eyewear whenever using compressed air to prevent eye injuries caused by air-blown particles.

Proper inflation is perhaps the most important part of tire service. Remember:

- **Air inflation equipment should have a filter to remove moisture.** This prevents corrosion inside the tire. Moisture may also deteriorate cord fabric of the tire and may result in eventual failure of the tire.
- **Always check and inflate when tires are cold.** This is very important since, as the tires roll and heat up, the air expands and the pressure increases. Automotive tire pressure may increase 4 to 6 psi (28 to 41 kPa). In truck and industrial tires, the pressure may increase even more.

NOTE: In some industrial tires it may take as long as 24 hours or longer for the tire to return to normal temperature.

- **Never "bleed" pressure from hot tires.** This invariably results in the tire pressure being too low when its temperature becomes normal. Either reduce the load, the speed, or both.
- **If you notice a low tire while operating, add air so the pressure is the same as the tire on the other side of the vehicle.** Recheck the pressure after about 30 minutes of operation.
- **Even when radial tires are correctly inflated, they may appear to be under inflated.** Always check the inflation pressure with an accurate tire gauge to prevent over inflation.
- **Always use a liquid-type pressure gauge when checking pressure in tires with liquid ballast. Check the pressure with the valve stem at the bottom.** Always wash the gauge with clean water after checking the tires.

NOTE: If it is impossible or undesirable to check the pressure with the valve stem at the bottom, locate the stem at the top. Add about 1/2 pound (3.4 kPa) to the pressure gauge reading for each foot height (30 cm) of the rim.

CAUTION: Failure to follow proper procedures when mounting a tire on a wheel or rim can produce an explosion which may result in serious injury or death. Do not attempt to mount a tire unless you have the proper equipment and expertise to perform the job.

When seating tire beads on rims or wheels, never exceed maximum inflation pressure specified by tire manufacturers for mounting tires. Inflation beyond this maximum pressure may break the bead, or even the rim, with dangerous explosive force.

INFLATING TRACTOR AND IMPLEMENT TIRES

Farm tractor and implement tires operate most of the time in field conditions where the lugs can penetrate the soil, and where all the width of the tread contacts the ground. When operating over hard-surfaced roads, with low inflation pressures, the tread bars squirm excessively while going under and coming out from under the load. On highly abrasive or hard surfaces, this action wears them down.

If the tires are to operate for any length of time on roads or hard surfaces and the draft load is light, increase the pressure to the maximum recommendation to reduce tread wear.

For a given size tire, the inflation pressure determines how much load the tire can carry. The proper inflation pressure for each wheel position should be based on the actual tire loads. The best way to determine the load being carried by the tires is to weigh the machine one axle at a time. Divide the total amount of weight on the axle by the number of tires on the axle to determine the weight being carried by each tire.

Use the tire inflation guidelines developed by the American Society of Agricultural Engineers (ASAE) to determine the proper inflation pressure. See your tire dealer. Besides using the ASAE guidelines, it is also a good idea to check the operator's manual for the proper tire inflation pressures, which sometimes vary with different operating conditions.

CAUTION: Many drive wheel rims, particularly on older tractors, are 0.187 inch (4.75 mm) thick and designed for a maximum inflation pressure of 24 psi (165 kPa). To avoid possible rim failure in operation, do not exceed this inflation pressure.

Ballast

Modern off-the-road machines are designed with adequate horsepower for their field or jobsite operation. However, the machine weight by itself may not be sufficient for full traction and drawbar pull, resulting in tire slippage. The effective weight on the driving wheels determines how much a machine can pull, depending on the ground surface.

The effective weight is the total weight of the machine (including added ballast) plus the weight that is transferred to the machine by an implement.

Since there is generally little control over the operating surface, tire slippage must be controlled by adding weight to the machine. Agricultural tractors are the prime users of wheel ballast.

TIRE TRACTION

Some modern tractors are equipped with a performance monitor that enables the operator to be constantly aware of the amount of wheel slip that occurs during high draft conditions. The tire traction can also be judged by looking at the tire tread pattern produced in the soil when pulling a heavy load.

TOO MUCH WEIGHT

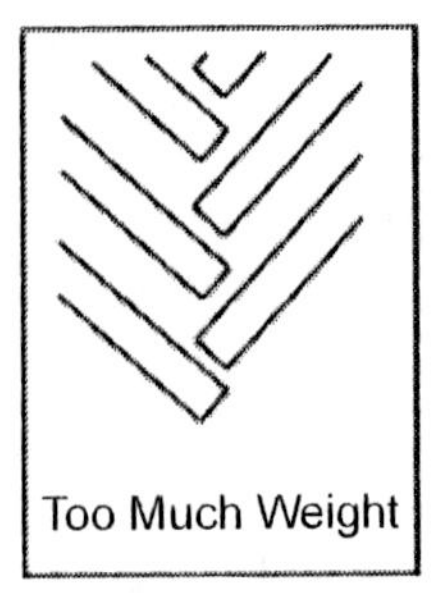

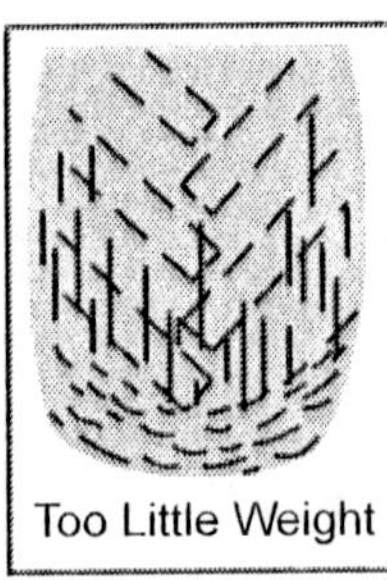

Proper Weight

JDPX1155

Fig. 18 — Tire Traction Patterns (Farm Tractor in Loose Soil Shown)

When *too much weight* is used, the tire tracks will be sharp and distinct in the soil (Fig. 18). There will be no evidence of slipping. Although this may seem to be ideal, it is bad, as the tires are literally geared to the ground and do not allow the flexibility that is obtained when some slipping occurs. With no wheel slippage, the shock loads that the tires would normally absorb are transferred to the power train, which can result in serious damage.

TOO LITTLE WEIGHT

When the tires have too little weight, they lose traction. The tread marks are entirely wiped out (Fig. 18) and forward progress is slowed. Not only is less work done, but the tires wear fast.

PROPER WEIGHT

When the tires have proper weight, a small amount of slippage occurs (Fig. 18). Usually, between 10 and 15 percent slippage is considered ideal in the field. When the tire is properly weighted, the soil between the cleats will be shifted but the tread pattern is still visible in the tire track. Proper weighting allows the engine to perform at its best with maximum flexibility.

WHY BALLAST?

Tractor tires develop pull in relation to the amount of weight upon them. The greater the weight, the more tractive effort the tire can exert. However, weight compacts soil and increases the rolling resistance. The heavier the tractor, the more power it needs to propel itself across the soil surface. There is, therefore, an optimum weight for a given tractor under given conditions. The use of proper ballast is normally the best way to obtain this optimum weight.

HOW MUCH BALLAST?

The amount of ballast which should be used depends on these factors:

- Soil Surface
- Type of Implement
- Travel Speed
- Tractor Power Output
- Tires

Soil Surface

On loose or sandy soil, more weight is required to develop full power than on a firm surface.

The surfaces referred to as firm to tilled (loose) are generally soil surfaces. A firm surface might be an alfalfa crop or corn stubble. A tilled surface might be a plowed or disked field. These cover a broad range of specific soil conditions.

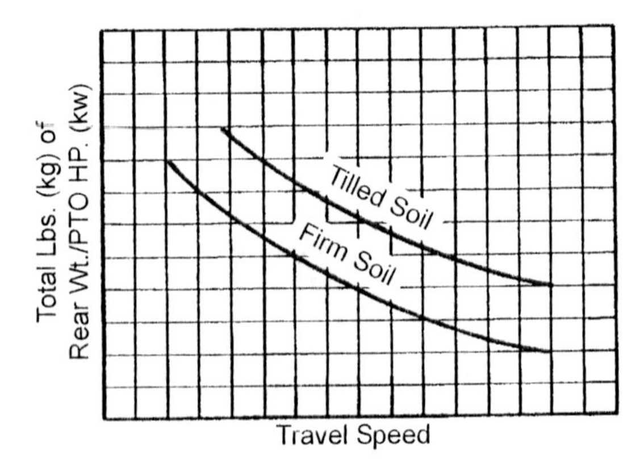

JDPX1156

Fig. 19 — More Weight Is Required to Develop Full Speed and Horsepower in Loose or Tilled Soil

More weight is required on a loose or tilled surface than on a firm surface for a given travel speed and engine power (Fig. 19).

Type of Implement

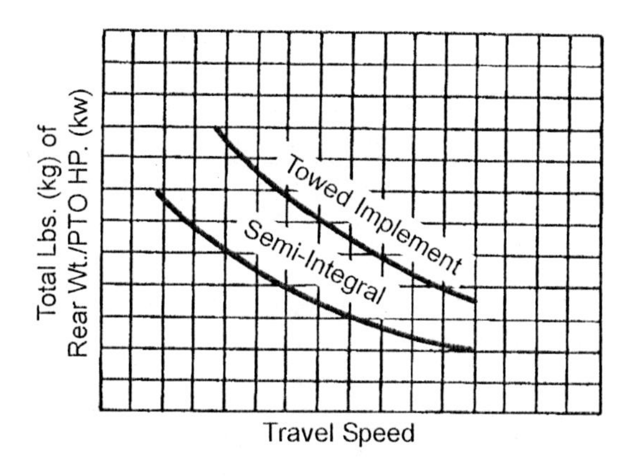

Fig. 20 — Less Weight Is Needed for Semi-Integral Implements

Integral and semi-integral (rear-mounted) implements transfer more weight to the rear wheels of a tractor than towed implements. Thus, less ballast is required for integral or semi-integral (rear-mounted) implements than for towed implements (Fig. 20).

Travel Speed

Less weight is required at higher travel speeds than at lower speeds. Figs. 19 and 20 both show decreasing rear weight as the speed increases.

Tractor Power Output

At a given travel speed, the weight required is in proportion to the PTO power output of the engine. Less weight is required for an 80-horsepower (60 kw) than for a 100-horsepower (75 kw) engine. Less weight is also required for a tractor that is not operating at full power. A tractor operating at 3/4 load requires only 3/4 as much rear weight as the same tractor operating at full load.

Tires

Tires have perhaps the least effect on ballasting requirements. However, better traction is usually obtained from over-size, dual or triple tires. A tire operating at its full capacity cannot be expected to perform as well as one that is used conservatively.

A highly ballasted tire will also operate with the greatest ground contact pressure and cause more soil compaction.

Radial tire construction contributes to better traction. The combination of flexible sidewalls and tough belting beneath the tread of a radial tire allows the weight of the tractor to press more of the tire into contact with the soil for a larger footprint.

Because the radial tire footprint is approximately 20 percent larger than a comparable bias ply tire, the radial tire transfers more power to the ground. This causes the tractor to be propelled forward with less slippage, increasing the tractor's field efficiency and fuel efficiency.

Dual tires can be used to reduce unit ground pressure to help avoid excessive soil compaction. They are also used when the carrying capacity of single tires is exceeded. However, adding dual tires does not necessarily mean that the rear weight must be increased.

HOW MUCH WHEEL SLIP?

Weight guides tell the amount of tractor weight required for a specific tractor under given conditions. However, they only give an estimate. The final criteria for adding ballast is the amount of travel reduction (% slip) of the drive wheels.

Under normal field conditions, travel reduction should be around 10-15 percent. Add more weight to the drive wheels if they still slip. If there is less than 10 percent slip, weight should be removed.

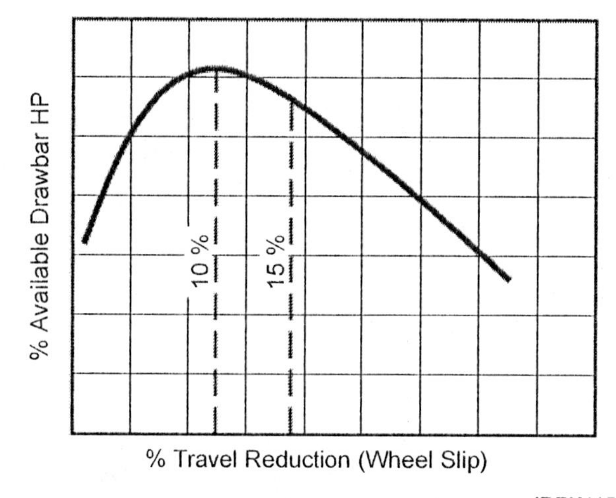

Fig. 21 — Available Horsepower Is Reduced when Wheel Slip Is Too Much or Too Little

Zero slip is not good. Fig. 21 shows what happens to the available drawbar horsepower when too much or too little slip is allowed. Note that the available horsepower is greatly reduced when wheel slip drops below 10 percent.

HOW TO MEASURE TRAVEL REDUCTION (WHEEL SLIP)

For a tractor without a performance monitor, the travel reduction or wheel slip can easily be found by the following method:

1. With chalk, make a reference mark on the side of one of the drive tires. Then, have someone drive the tractor in a

straight line while under full implement load at the normal speed and gear range to be used.

2. Walk alongside the tractor, making sure you are clear of the implement. Mark a starting reference point on the ground where the chalk mark first comes down to the ground with a stick, flag, etc.

3. Walk along the side as the tractor moves through the field under full implement load and count off ten revolutions of the wheel. Mark the spot where the chalk mark comes down to the ground the tenth time.

4. Take the implement out of the ground and return to the starting point. Mark the tire where it comes down on the first mark previously made on the ground. Count the number of revolutions of the wheel required to travel the distance between the two marks on the ground with the implement out of the ground, estimating the last revolution as closely as possible.

5. Determine the slip as follows:

Revolutions	% Slip	Result
10	0	Remove Ballast
9-1/2	5	
9	10	Proper Ballast
8-1/2	15	
8	20	Add Ballast
7-1/2	25	
7	30	

If less than 8-1/2 revolutions are obtained, *weight should be added*. If more than 9 revolutions are obtained, *weight should be removed* from the rear wheels.

NOTE: The formula for calculating wheel slip is as follows:

R^1 = Revolutions with no load over a given distance

R^2 = Revolutions with load over the same distance

$$\text{Percent Slip} = \frac{R^2 - R^1}{R^2}$$

FRONT BALLAST (FOR TWO-WHEEL DRIVE MACHINES)

Front ballast does not contribute directly to traction on two-wheel drives but is required for stability and safety. Approximate front weight should be as follows:

For integral or semi-integral implements —
Front weight = 1/3 rear weight

For towed equipment attached to the drawbar —
Front weight = 1/4 rear weight

NOTE: The front weight should be increased more if necessary for safe operation of the tractor. Check the operator's manual charts.

FRONT BALLAST (FOR FRONT-WHEEL DRIVE MACHINES)

Tractors equipped with a mechanical front-wheel-assist drive axle will achieve optimum operating efficiency when the static rotor weight split is set at approximately 40 to 45 percent on the front axle and 60 to 55 percent on the rear axle.

The weight distribution for four-wheel drive tractors should be approximately 55 to 60 percent on the front axle and 45 to 40 percent on the rear axle.

To achieve the correct static tractor weight split on the axles, use cast iron weights on the rear of the tractor and a combination of liquid ballast and cast iron weights on the front of the tractor.

POWER HOP

Under high drawbar loads, in certain soil conditions, tractors with a mechanical front-wheel-assist drive axle and four-wheel drive tractors may simultaneously experience a loss of traction and a bouncing and pitching ride. The occurrence of these two conditions at the same time is termed "power hop."

In some instances, the power hop may become so severe that the operator loses the ability to safely control the tractor. To regain control, the operator should reduce the engine speed and/or reduce the implement draft until the power hop subsides.

One misconception is that the use of bias ply tires will cure power hop. This is not true. Radial tires have less stiffness and damping effect than bias ply tires, but tractors equipped with either type tire can experience power hop.

Research has shown that power hop can usually be controlled by properly ballasting the tractor to obtain the recommended front/rear weight ratios, and by changing the tire inflation pressure to alter the tire deflection. Also, match the implement size to the tractor power, be sure the implement is correctly adjusted, and do not exceed the optimum implement operating speed.

To control power hop on a properly ballasted front-wheel drive tractor, the front tires must be stiffened and the rear tires softened.

IMPORTANT: Liquid ballast in the tires has a dramatic stiffening effect at low inflation pressures; therefore, liquid ballast in the rear tires should be avoided when a power hop condition exists. Filling the front tires with fluid reduces the tire deflection and may eliminate hop.

Proceed as follows to alter the inflation pressure in the front and rear tires to increase the stiffness of the front tires and decrease the stiffness of the rear tires. First, increase the front tire inflation pressure to the maximum recommended pressure for the particular tire size, ply rating and static load.

For single rear tires, set the tire inflation pressure to the minimum pressure shown in the tire inflation chart in the tractor's operator's manual for the particular tire size, ply rating and static load. If dual rear tires are used, decrease the rear tire inflation pressure to 14 psi (100 kPa) on the inner drive tires and to 12 psi (80 kPa) on the outer drive tires.

NOTE: Depending on the weight the tires must support, it may be possible to reduce the rear tire inflation pressure below 12 psi (80 kPa) if radial drive tires are used. Refer to the extended tire inflation tables for radial drive tires (Fig. 12).

BALLAST LIMITATIONS

Ballast should be limited by either the tire capacity or tractor capacity. Each tire has a recommended carrying capacity which should not be exceeded. If a greater amount of weight is needed for traction than is allowable on the tire, either a larger single tire should be used or duals should be considered. Tractor and tire life can be extended if ballast is limited to that required for continuous operation in the **higher** range of speeds, usually 4 mph (6.4 km/h) or above.

Since ballast is usually added to a tractor and seldom removed for light loads or higher speed operation, a compromise is usually required. If the tractor is used mostly for high draft loads at low speeds, ballast for this type of operation. If more time is spent on light load utility work or at higher speeds, ballast for this operation and allow more slip for the smaller time spent on high draft work.

TRACTOR WEIGHT GUIDES

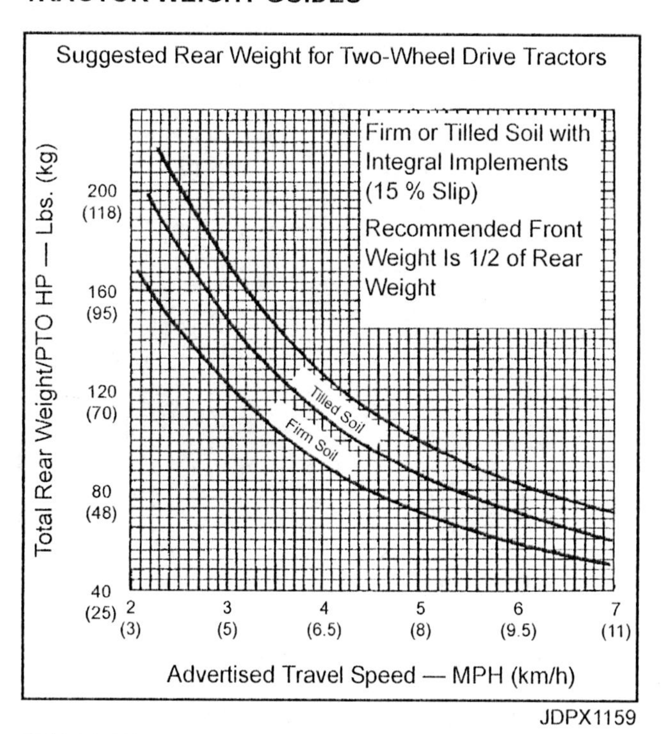

Fig. 22 — Tractor-Weight Guide for Integral Implements

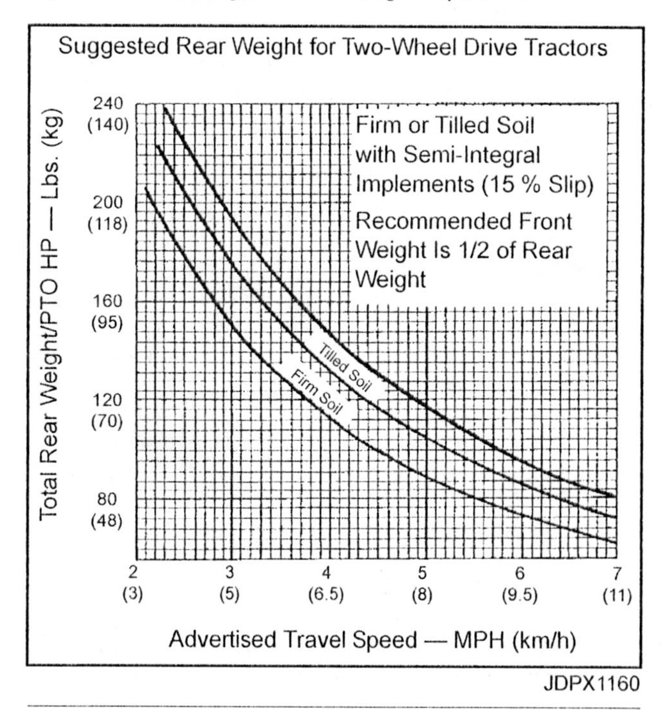

Fig. 23 — Tractor-Weight Guide for Semi-Integral Implements

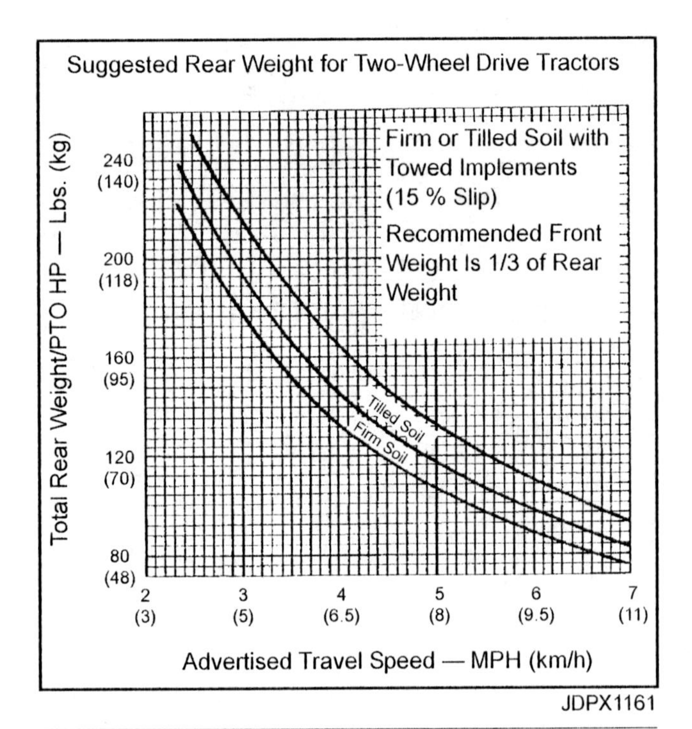

Fig. 24 — Tractor-Weight Guide for Towed Implements

The tractor weight guides illustrated in Figs. 22, 23 and 24 can be used to determine the amount of ballast to be added. The guides indicate the pounds (kg) per PTO (power take off) power of rear tractor weight under various soil and implement conditions.

Since the tables give ballast in terms of total tractor rear weight, you need to know the actual rear weight of the tractor as a basis to start. Find the actual rear weight as follows:

1. Drive the tractor on a platform scale, leaving the rear wheels on the scale and the center of the tractor at the edge of the platform.

2. Be sure the tractor is as level as possible, blocking under the front wheels if necessary.

3. Place the transmission shift lever in neutral.

4. Obtain the weight.

NOTE: The front end weight of the tractor can be found the same way, placing the front wheels on the scale.

The charts also tell the amount of recommended front tractor weight in relation to the rear tractor weight. As stated earlier, adding weights to the front of a two-wheel drive tractor is a matter of stability and safety.

The following example explains the use of the Tractor Weight Guide.

A tractor is to be used with a semi-integral plow on a grass cover crop. The tractor is to be operated in 4th speed which is 5.3 miles per hour (8.5 km/h).

In the Fig. 23 chart on "Semi-Integral Implements," find the range of Total Rear Weight/PTO HP—Lbs. (kg) at 5.3 mph (8.5 km/h) on *firm* soil. From the chart, this range is between 84 and 100 lbs. (51 and 60 kg).

If the tractor is to be operated at full load, 120 PTO hp (90 kw) for our example, the rear tractor weight should be between 10,100 and 12,000 lbs. (4590 and 5400 kg) as calculated below.

120 PTO hp x 84 lbs./rear PTO hp = 10,100 lbs.
(90 PTO kw x 51 kg/rear PTO kw = 4590 kg)

120 PTO hp x 100 lbs./rear PTO hp = 12,000 lbs.
(90 PTO kw x 60 kg/rear PTO kw = 5400 kg)

If the middle of the firm soil range is used, the rear weights for full load operation would be about 11,000 lbs. (4950 kg).

However, if you want to operate the tractor at 3/4 load, 90 hp (67 kw), the rear weight of the tractor should be within the range below:

90 x 84 = 7,550 lbs.
(67 x 51 = 3417 kg)

90 x 100 = 9,000 lbs.
(67 x 60 = 4020 kg)

If the middle of the firm soil range is used, the rear weights for 3/4 load operation would be about 8,300 lbs. (3725 kg).

From these figures, and the unballasted front weight, you can also calculate the amount of weight that must be added to the front.

CAST IRON AND LIQUID BALLAST

Tractor Tire with Liquid
in Tube as Wheel Ballast

Tractor Tire with Removable
Cast-Iron Wheel Ballast

JDPX1162

Fig. 25 — Liquid and Cast Iron Ballast on Drive Wheels

Ballast can be added as either liquid or cast iron (Fig. 25). Filling the tires with liquid ballast, a mixture of water and calcium chloride to prevent freezing, is a job for a tractor dealer or tire service store. Because of the special equipment needed for adding liquid ballast to the tractor tires, it is often regarded as a permanent part of the tractor and not removed when the tractor is used for light draft jobs.

Cast iron weights are more readily installed and removed. Thus, the general practice is to install liquid ballast permanently in the tires, and then use cast iron weights as the adjustable portion of the total ballast.

Cast Iron Weights

JDPX1163

Fig. 26 — Ballast a Farm Tractor for Operation in Higher Gears — Usually 4 mph (6.4 km/h) or Above

JDPX1164

Fig. 27 — Cast Iron Front End Frame Weights on a Farm Tractor

Cast iron ballast weights can be obtained from the tractor manufacturer. They weigh 40 to 140 pounds (18 to 64 kg), although heavier weights are often installed at the factories as standard equipment. Cast iron ballast can be combinations of wheel weights (Fig. 26) and frame weights (Fig. 27). One advantage of cast iron weights is that they can be added or removed to suit the operating conditions.

Wheel weights are provided in a variety of sizes, which enables the tractor owner to finely adjust the tractor's weight and weight distribution. The ease of installing and removing wheel weights can vary widely. For instance, weights up to 100 pounds can be installed by one person, or preferably two, if a provision is made for a knob or ledge in the weight to support the next weight, or if a provision is made for the attaching bolts to be held solid while the next weight is slid into place.

CAUTION: Always use the proper procedures and body position when lifting to reduce the risk of injury. Protect your hands and feet with gloves and steel-toed safety shoes. Keep your fingers and feet clear of pinch points to avoid injury.

If the weights must be supported in an exact location while the attaching bolts are inserted or while the retaining nuts are tightened, two people or a hoist are needed to safely install the weights.

Wheel weights weighing more than 100 pounds (45 kg) should be lifted and positioned with a forklift or hoist.

Liquid Ballast

Liquid ballast, usually water, may be added to both the front and rear tires to increase the load on the drive axle. Calcium chloride is normally added to the water. Adding calcium chloride to the water offers the following advantages:

- Additional weight (up to 50 percent) over plain water.
- It is not harmful to the tire.
- It is plentiful and low in cost.
- It is an effective antifreeze solution.

Tubeless rims can be corroded by the calcium chloride solution if they aren't completely covered with fluid when the tractor is parked. For this reason, some operators fill tubeless tires with a water and ethylene glycol antifreeze solution to avoid rim damage. However, the ethylene glycol solution is about equai in weight to plain water.

There are many methods of installing liquid ballast, but a tractor dealer or tire dealer normally has the equipment and the knowledge to do the work.

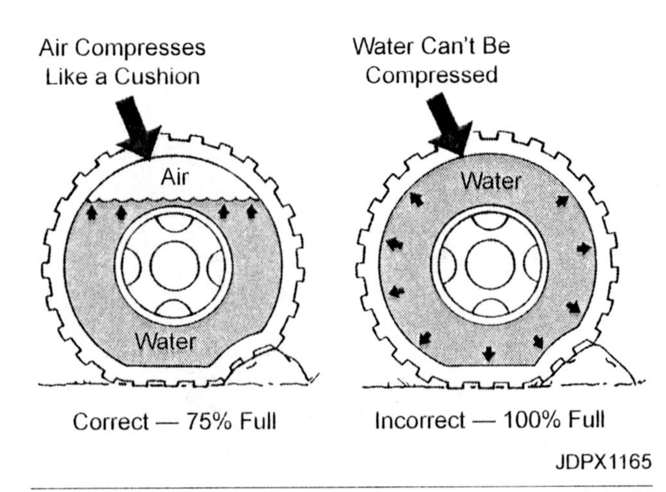

Fig. 28 — Why Tires Should Be Filled Only 75 Percent Full of Liquid Ballast

When filling a tire with liquid ballast, **fill it only 75 percent; or up to valve level** (Fig. 28). This is recommended for the following reasons:

- An air chamber is needed to retain the pneumatic principle.
- Water or liquid solutions cannot be compressed (the hydraulic principle). If a tire is filled with liquid, it has no air chamber to absorb shocks, and has little or no bruise resistance.

Adding Liquid Ballast

LIQUID BALLAST CHART

Tire Size (Drive Wheels)	Ply Rating	Max. Carrying Capacity Lbs.* (kg)	Volume of Liquid Gal (L)	Weight of Liquid Including $CaCl_2$** Lbs. (kg)
13.6 - 28	4	4860 (2206)	74 (280)	878 (399)
13.6 - 38	6	6940 (3151)	98 (371)	1162 (529)
14.9 - 28	6	6640 (3015)	92 (348)	1090 (496)
15.5 - 38	6	7320 (3323)	112 (424)	1326 (603)
15.5 - 38	8	8660 (3932)	112 (424)	1326 (603)
16.9 - 34	6	8340 (3786)	140 (530)	1658 (754)
16.9 - 34	8	10040 (4558)	140 (530)	1658 (754)
16.9 - 38	8	10640 (4831)	154 (583)	1824 (830)
18.4 - 34	6	9880 (4486)	170 (643)	2014 (916)
18.4 - 34	8	11300 (5130)	170 (643)	2014 (916)
18.4 - 38	8	11940 (5421)	188 (712)	2226 (1013)
18.4 - 38	10	13900 (6311)	188 (712)	2226 (1013)
20.8 - 38	10	15300 (6946)	240 (908)	2842 (1293)
24.5 - 32	10	17800 (8081)	292 (1105)	3458 (1573)

* Capacity values are for TWO single drive wheel tires.

** Liquid ballast weight is based on water with 3-1/2 Lbs. (1-1/2 kg) of calcium chloride ($CaCl_2$) per gallon (3-3/4 L) and tires filled to valve level (75 percent full). This mixture protects against freezing (slush-free) to –12°F.

The Liquid Ballast Chart lists **maximum** amounts of liquid ballast for drive wheel tires. The maximum carrying capacity for two single tires is also listed. *Never exceed this carrying capacity.* For the actual recommended ballast to carry, always refer to the machine operator's manual.

DRY BALLAST

Dry powder ballast is sometimes used for weighting tires. Some dealers are equipped to install dry ballast, and a few manufacturers install it at the factory. Normally, tires are filled 87 to 95 percent full.

Advantages claimed with the use of dry ballast are:

- Gives 20 to 30 percent more traction with less tire wear.
- Reduces the inherent bounce in rubber tires, making the machine smoother and easier to operate.
- There is no chance of freezing.

Dry ballast is furnished in three different weights: 10, 15, and 20 pounds per gallon (4.5, 6.8, 9 kg). The 15-pound (6.8 kg) weight is recommended for 90-95 percent of all farm tractors, and the 20-pound (9-kg) weight for heavy equipment.

There are some **disadvantages** to dry ballast:

- Higher initial cost than other types of ballast.
- Requires servicing with special equipment.
- You lose the ballast if there is a tire, tube, or valve failure.

In summary, dry ballast has shown its potential and some manufacturers now recommend it for use with their equipment.

Dual and Triple Tires

JDPX1166

Fig. 29 — Dual Tires

As the engine horsepower in a rubber-tired tractor is increased, traction problems also increase. The use of dual tires (Fig. 29), or even triple tires, is one way to add more carrying capacity. This in turn allows more weight to be applied through the use of ballast (liquid, dry, or metal), resulting in greater traction.

ADVANTAGES OF DUAL OR TRIPLE TIRES

- Dual or triple tires, with added ballast, give you more ground contact. Result: Less slippage, increased ground speed, improved work rates, and possibly reduced fuel consumption.
- Improved tractor stability.

- Dual or triple tires lessen driver fatigue by providing a smoother ride.
- Dual or triple tires give more flotation, and help keep the tractor on top of the ground.
- Tractors which can be converted to duals or triples have a higher potential for all-season use. Duals or triples can be used to meet the higher horsepower requirements for land preparation.
- Duals or triples permit use of the tractor under many adverse weather and soil conditions not feasible when using singles. Result: Plowing and disking can start earlier in the spring.
- Duals or triples permit a tractor with one tire requiring service to be moved to a convenient place for repairs, thus reducing downtime.

DISADVANTAGES OF DUAL OR TRIPLE TIRES

- Some tractor axles, bearings, and power trains may be overstressed by dual or triple wheels. Always contact your local tractor and implement dealer for advice before installing duals or triples.
- Reducing the air pressure in dual or triple tires, for riding comfort, may affect tire life by causing overdeflection of tires. Consult the operator's manual or the local tire dealer for recommended air pressures.
- If the drawbar load is fairly light, there is no real advantage in dual or triple tires over single tires of the same size, other than riding comfort, better stability, and less soil compaction.
- Dual or triple tires may make a tractor more difficult to turn, and if short turns are made, the extreme stress on the tread lugs may tear the lugs.
- Changing a large dual or triple tire may be difficult unless you have the proper equipment. Specialized equipment for changing large farm tires is usually available at farm tire dealers.

MATCHING DUAL OR TRIPLE TIRES

Mismatched dual or triple tires cause uneven distribution of the load, which makes the larger tire carry too much load and wear faster. To avoid this, tires of approximately the same diameters or overall height should be used.

There should not be more than 1/4 inch (6 mm) difference in diameter or 3/4 inch (19 mm) circumference on 8.25 or smaller tires; and not more than 1/2 inch (13 mm) in diameter or 1-1/2 inches (38 mm) in circumference on sizes larger than 8.25. The most accurate check is to take a circumferential measurement with a steel tape on tires mounted and inflated.

Tandem drive machines should have matched tires on all drive wheels. Failure to keep tires uniform in size will result in excessive slippage and fast wear of the odd-sized tires. It may also break an axle, make steering difficult, and cause other problems.

How to Check Sizes of Dual Tires Mounted on a Machine

METHOD NO. 1

JDPX1167

Fig. 30 — Checking Sizes of Dual Tires with a Square

Use a right-angled square made of 1 x 2 inch (25 x 51 mm) wood strips to measure the diameter differential of dual tires (Fig. 30). This measuring tool should be squared with a carpenter's steel square and rigidly fastened to maintain a true 90 degree angle. If one tire is too small, it becomes apparent at once as the wood strip is laid across the tires.

METHOD NO. 2

JDPX1168

Fig. 31 — Checking Sizes of Dual Wheels with a Cord

Use a cord (Fig. 31) as a quick and easy way of checking duals. A cord can be easily improvised using two rubber bands cut from old tubes, two hooks made from a welding rod, and a length of 1/8 inch (3 mm) cord. Hooked to a wheel stud or valve, and drawn across the tires, the cord quickly shows if the inside tire is smaller. If the cord touches both tires, any difference in diameter can be determined by slowly lifting the cord from the outside tire until it is just touching the inside tire.

ALTERNATIVES TO DUAL TIRES

There are many tires produced for off-the-road service. Some of these are made extra wide to be used in place of dual tires. Others come as standard equipment on earth-moving machines and the like. Still others are installed to replace dual tires.

Before replacing duals with these tires or vice versa, consider these factors:

- Load-carrying capacity
- Adaptability to the machine
- Cost

As a rule, a manufacturer installs tires which are most practical and safe for his product. He is as interested as you are in the machine giving the best service, for the longest time, for the least money.

Front Tractor Tires and Implement Tires

Like rear drive tires, front tractor and implement tires have a large variety of treads, sizes, and ply ratings. Some, which are used to provide power to lift a plow out of the ground or drive some component on the machine, have cleats similar to tractor drive tires. However, most front tires have ribbed treads for good steering control.

Specially designed front tires for use on front-wheel assist tractors are also available. These tires better resist the high side-rolling stresses generated during sharp turns when operating a front-wheel assist tractor. These tires are also designed to better withstand the wear and scuffing that is a result of the faster speed of the front-wheel assist tires compared to the rear tires.

Switching Tire Types and Sizes

Replacement tires should be the same type, size, and ply rating as original tires.

In some cases, such as unusually severe service or special traction or flotation conditions, you may consider a different size, type, or stronger tire.

Always bear in mind, that a tire with low initial cost is not necessarily the least expensive in the long run. The tire that will deliver maximum performance over a long period of time frequently costs less when you calculate on a basis of cost-per-hour of operation or work performed in similar service.

The performance of the tire being replaced, service conditions, terrain, loads, and dimensional limitations must be considered. Reference data which permits proper matching of tire to vehicle and type of service is an everyday tool of any reputable tire dealer.

We recommend you consult such a tire dealer for professional advice when selecting replacement tires.

Tire Failures

In this section we will cover the common tire failures and their causes. In the next section on "Tire Repairs" we will look at examples of how the service person repairs tires.

FABRIC BREAKS

JDPX1169

Fig. 32 — This Typical Small Fabric Break Can Be Repaired and the Tire Kept for Its Full Service Life.

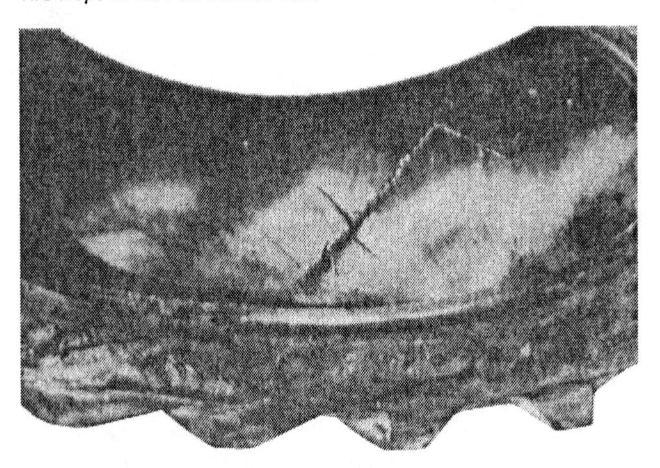

JDPX1170

Fig. 33 — Break Caused by Tire Hitting an Object. Fabric was Unable to Withstand the Shock.

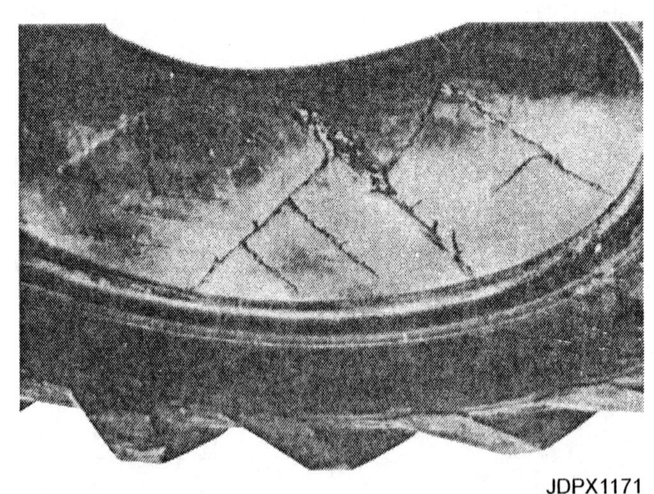

JDPX1171

Fig. 34 — Severe Rupture of Cords Extending from Bead to Bead, Caused by an Impact at High Speed or with Overinflated Tire.

Most fabric breaks are caused by hitting some object which puts too much shock on the fabric (Figs. 33 and 34).

The force is greater if the impact occurs at high speed, or if the tires are overinflated (Fig. 34). At high inflation pressures, injuries may occur which later result in a large "X" or diagonal break which may extend from bead to bead.

Even at correct tire inflation, a severe localized blow, such as hitting a sharp rock or tree stump, can produce a penetrating force which may break cords.

To reduce the hazard of tire breaks, use extra ply tires, correct tire inflation, and drive carefully.

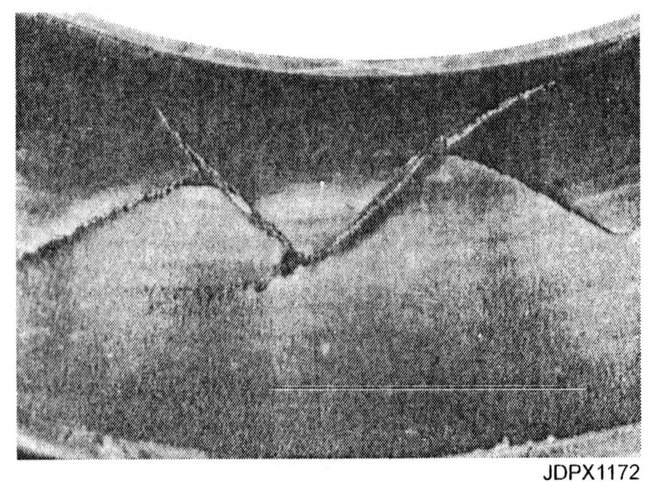

JDPX1172

Fig. 35 — Tire Ruined from Water Ballast Freezing and Expanding Inside the Tire.

A special kind of tire break is caused by water freezing and expanding in the tire (Fig. 35). To prevent this, always use an antifreeze solution ($CaCl_2$) as liquid ballast in your tires.

Fabric Breaks On Furrow Tires (Farm Tractors)

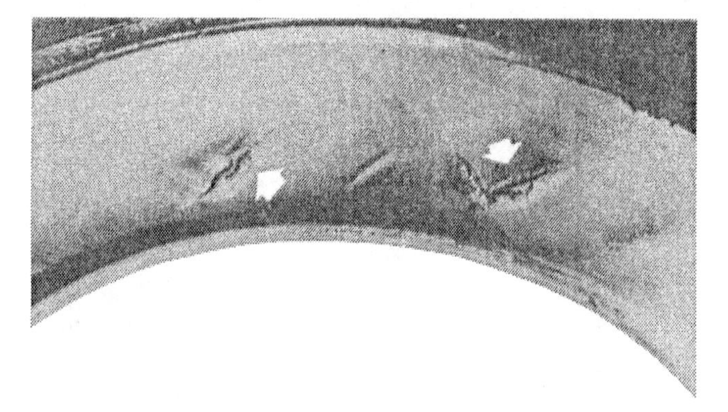

JDPX1173

Fig. 36 — Furrow Tire on Farm Tractor with Low-Pressure Buckle or Furrow Break

On farm tractors, the furrow tire may be distorted by the tilt of the tractor. The tilt folds over the tire, especially if the inflation pressure is too low.

The inner sidewall of the tire may buckle and fold until the cords separate and a series of breaks appear on the inside of the sidewall (Fig. 36).

JDPX1174

Fig. 37 — Furrow Tire with Sidewall Cracking from Underinflation and Heavy Drawbar Load

This action may also cause a series of cracks at the edge of the tread bars (Fig. 35). Sometimes the cracks may extend into the sidewall as shown.

To help prevent stress on the furrow tire, make these adjustments:

1. Place 4 pounds (28 kPa) more pressure in the furrow tire than in the land tire. Don't exceed the maximum recommended pressure.

2. Adjust the plow hitch laterally so that the furrow tire does not need to crowd the furrow wall to plow a full-width cut.

RUBBER CHECKS

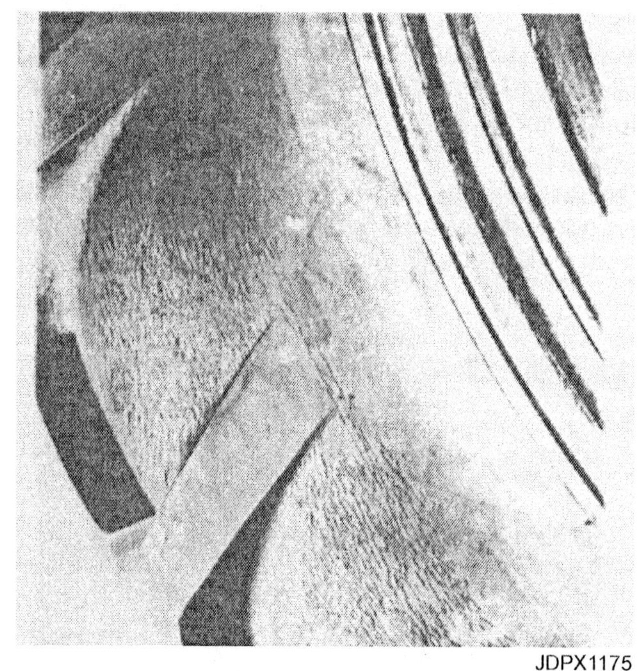

JDPX1175

Fig. 38 — Weather Checks on the Tire — Usually Not Harmful

Smaller checks or cracks may develop on the rubber sidewall (Fig. 38). These are usually caused by tires inflated to high pressures and exposure to sunlight, ozone, air drafts, or electrical discharges.

Normally this is an appearance condition only and will not affect the service life of the tires. Correct inflation pressure and protection from the elements will help prevent this condition. Even a good coating of mud on the tires may be helpful.

WEAR FROM SPINNING WHEELS

JDPX1176

Fig. 39 — Tire with Spinning Wear

Tractor tires with too few wheel weights or too much inflation pressure will wear the tread bars rough or will snag and cut the bars when subjected to severe service on abrasive surfaces. Sudden engagement of the clutch also causes this type of tread wear. Tread bars are cut and worn on the leading edge (Fig. 39).

To prevent spinning wear, add wheel weights, reduce inflation pressure to normal for conditions, reduce the draft load where possible, and engage the clutch slowly when starting.

STUBBLE WEAR

JDPX1177

Fig. 40 — Farm Tractor Tire with Stubble Wear

Stubble wear on the tread rubber is caused by operating a tractor tire over crop stubble (Fig. 40). Stubble can be rigid enough to puncture a tire. To avoid or limit stubble damage, adjust the track width of the rear tires so neither tire rides over the stubble, and don't spin the wheels when they do contact stubble.

HARD ROAD OPERATION

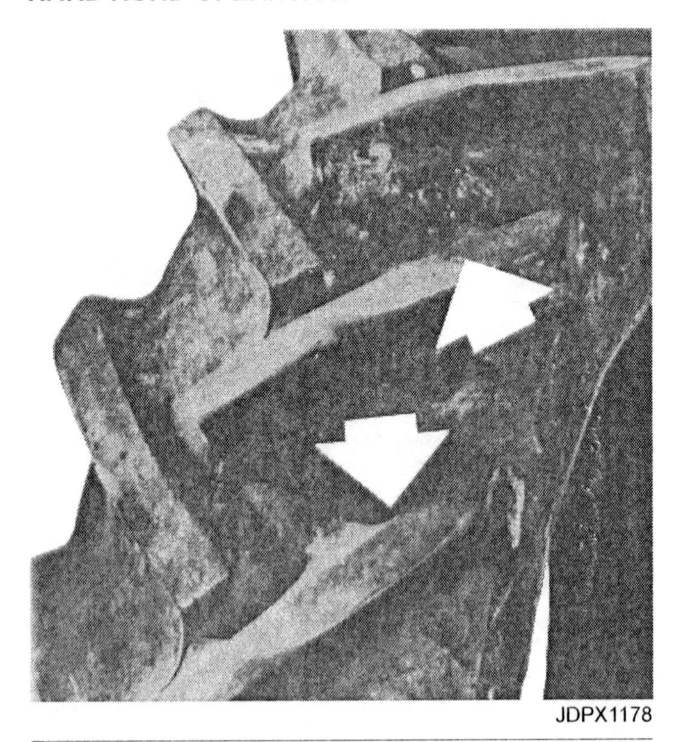

JDPX1178

Fig. 41 — Tread Wiping from Hard Road Operation

Tractor tires usually operate in the field where soil can conform to the tread design, and where all portions of the tread make contact with the soil. When you are on the hard road with low inflation pressure, there is an undesirable distortion of the tire which makes the tread bars squirm going under and coming out from under the load.

On highly abrasive, hard surfaces, this action wipes the rubber off the tread bars and wears them down prematurely and irregularly (Fig. 41).

If you plan to travel a long distance on hard surfaces with a light load, *increase the pressure in the tires to the maximum recommended* to reduce this tread movement. Under these conditions wear in the center of the tread can be expected.

Farm tractor and implement tires are designed for low-speed operations not exceeding 20 mph (32 km/h). If tractors and implements are towed at high speeds on the highway, heat may develop under the tread bars which will weaken the cord material and cord fabric. This may not be visible at the time, but may result in a premature failure.

VALVE DAMAGE

When valves are torn off tubes, it indicates a slippage of the tire bead on the rim or improper centering of the valve in the hole of the rim.

Slippage of the tire bead on the rim may be caused by:

- Low inflation pressure.
- Improper seating of the bead on the rim.
- Excessive use of soap solution, or using a petroleum base or silicone lubricant on the tire beads or rim when mounting the tire.

When the valve slips, the tire should be demounted and the rim and wheel carefully cleaned. Then the tire and tube should be mounted and inflated, deflated, and then reinflated as described under "Mounting Inner Tubes on Tube-Type Tires and Inflating" in a later section.

TREAD AND SIDEWALL CUTS

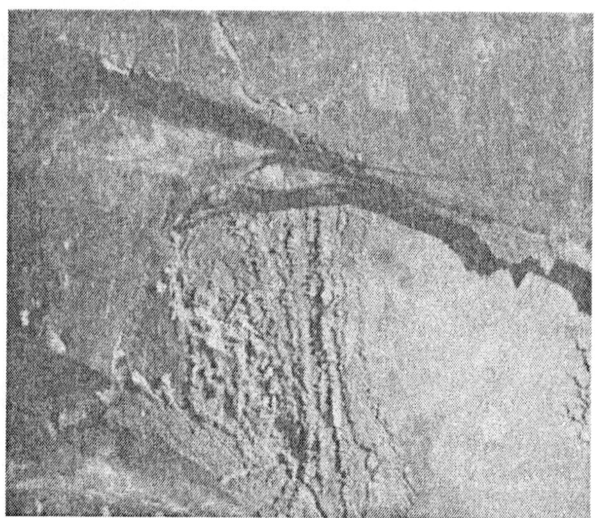

JDPX1179

Fig. 42 — Prompt Repair of Cut at Left Would Have Prevented Damage at Right

Most tires injured by cuts or snags can be repaired and continued in service. Inspect tires periodically for possible injuries. Cuts or breaks that enter into or expose the cords in the body of the tire should be promptly repaired. If this is not done, moisture and foreign material will enter the injury, deteriorate the cords, and make it impossible to restore the tire to service (Fig. 42).

A tire injured in this way should be removed promptly from the wheel and sent to a reputable tire service station or repair shop where a permanent repair can be vulcanized into the tire. If, in an emergency, it is necessary to continue to use the tire after it has been injured, a blowout patch or boot can be inserted at the point of injury to reinforce it. Never use a temporary repair any longer than three or four days before you remove the tire and have it repaired by a competent tire serviceman.

JDPX1180

Fig. 43 — Small Cut in Tire Lug (Left) Bevelled Out to Prevent Imbedded Stones (Right)

When a cut or snag extends only into a rubber lug and not into the tire cords (Fig. 43), cut out any loose piece of rubber with a knife so the tear won't grow and pick up stones. This cut should be made as smooth as possible by bevelling out the sides of the injury from the bottom to the surface of the lug as shown. Always consult your tire serviceman as soon as possible.

Repair of tires is given in detail in the next section on "Tire Repairs."

GREASE AND OIL DAMAGE

Keep grease and oil off tires, as both destroy rubber. After using a machine for spraying, wash off any chemicals that may have dropped on the tires. Even though the tread and sidewall rubber of off-the-road tires is compounded to resist sunlight, always keep the machine under cover and jacked up when not in use over a long period. Only when it is put back into service should the tires be inflated to correct pressure.

BELT WORK

When a rubber-tired machine is used to drive a belt, static electricity may develop. If the machine is not grounded, this may damage the tires. Ground the machine with a chain, wire, or rod from the metal frame to the ground.

TIRE OBSTRUCTIONS

A bent frame or fender may rub on the tires. Sharp corners can gouge the tire. Check for these tire obstructions often.

Tire Repairs

Both tube-type and tubeless tires can be repaired in much the same way. If tire injuries are detected early, they can usually be repaired before a blowout or other failure occurs.

FLAT TIRES

JDPX1181

Fig. 44 — Irreparable Tire Damage Resulting from Driving Flat.

If a tire loses all or most of its air pressure, particularly when driving at high speeds, it must be removed from the wheel for complete internal inspection to be sure it is not damaged. Inspections have shown that as many as 2 out of 3 tires, run even a short distance while flat, are damaged beyond repair (Fig. 44).

Punctures, nail holes, and cuts, up to 1/4 inch (6 mm), can be repaired permanently from the inside of the tire. Consult your tire supplier for information on larger repairs.

TEMPORARY REPAIR OF TIRES

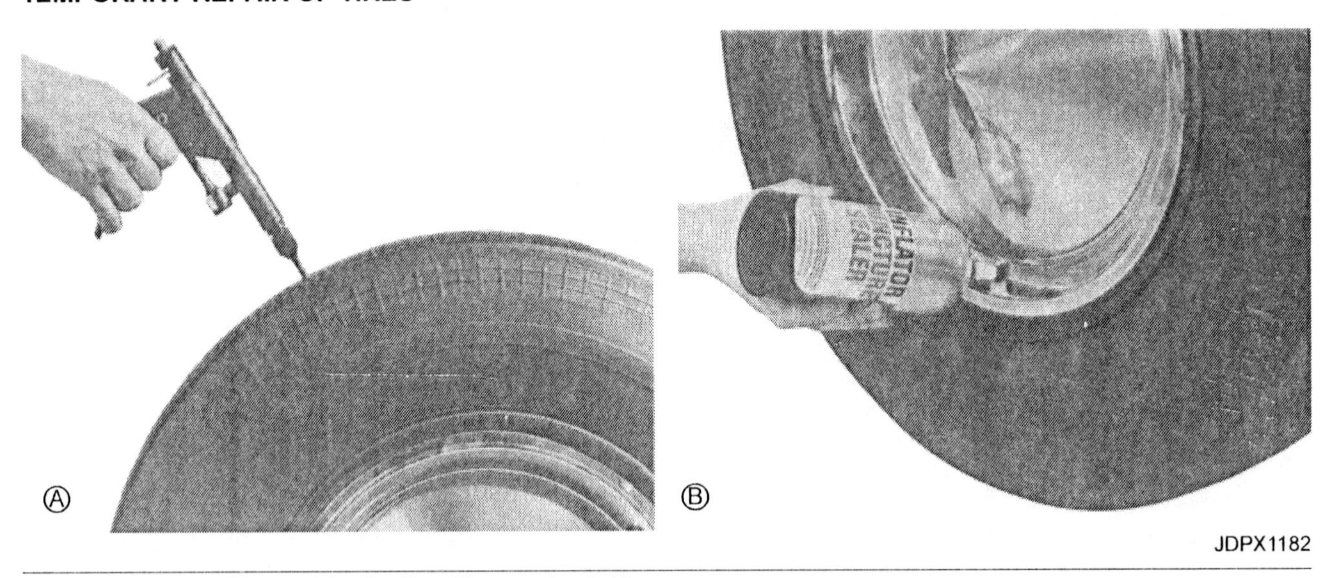

Fig. 45 — Temporary Repairs of Smaller Tires (Emergencies Only)

Temporary repairs, such as "blowout" patches or any outside tire repair, should not be made except in case of emergencies (Fig. 45). Emergency repairs, such as outside plugs (A) or aerosol sealants (B) are only good for a short time, and then only at low speeds. If these temporary repairs are used, make a permanent vulcanized repair with a plug and patch applied from the inside of the tire as soon as possible. In emergencies, a good spare tire, properly inflated, is the best insurance.

IMPORTANT: For radial ply tires, repairs can only be made in the central tread area, between the major outside grooves.

PERMANENT REPAIR OF TIRE PUNCTURES — GENERAL GUIDELINES

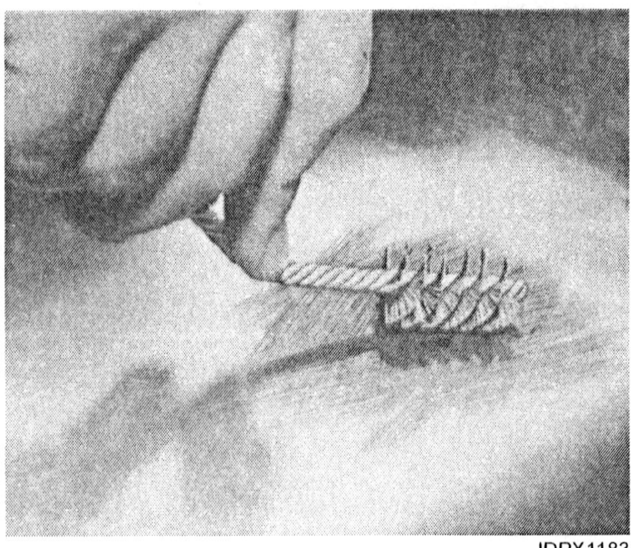

Fig. 46 — Preparing Injury for Repair

1. After removing the tire from the rim, probe the injury to remove nails or other damaging items. Make sure that the area around the injury is thoroughly dry. Scrape the damaged area with a sharp-edged tool and buff (Fig. 46). Be careful not to damage the liner or expose any cords.

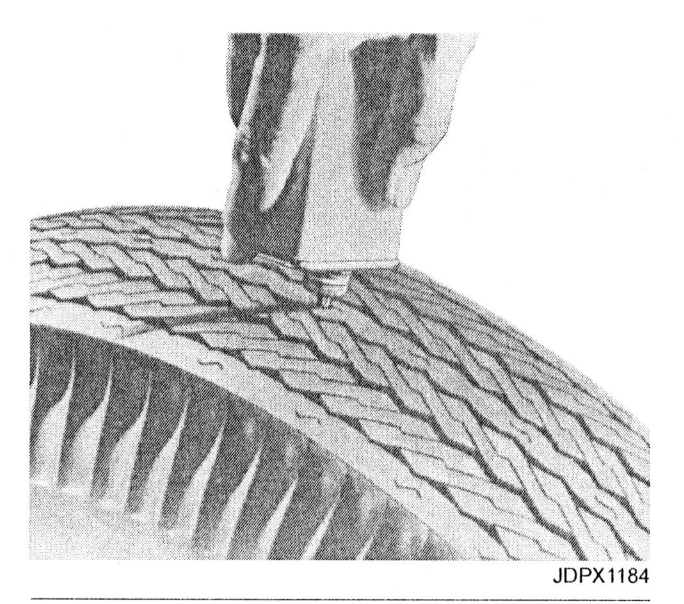
JDPX1184

Fig. 47 — Lubrication of Injury

2. Lubricate the injury by pushing the snout of the vulcanizing fluid can into the injury from both sides of the tire (Fig. 47). Also pour vulcanizing fluid on the insertion tool and push it through with a twisting motion until it can be inserted and withdrawn easily.

NOTE: Each patch or plug kit should contain specific instructions.

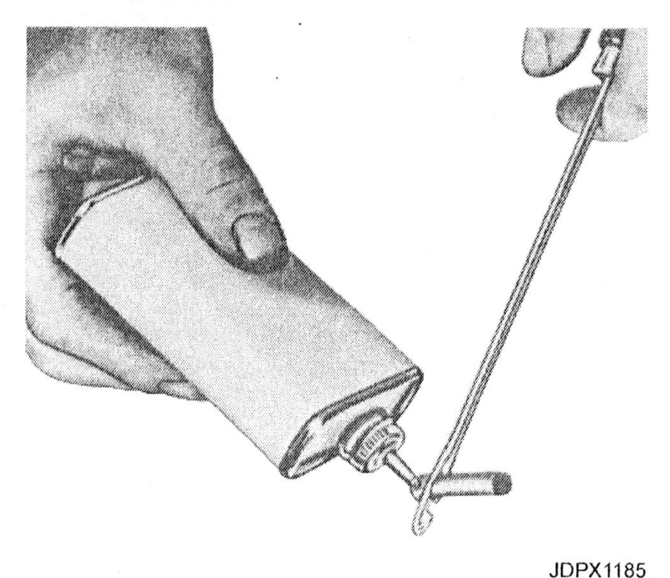
JDPX1185

Fig. 48 — Lubricating Tool and Plug

3. Using a head type or headless straight plug slightly larger than the size of the injury, place it in the eye of the insertion tool. When a headless straight plug is used, always back it up with a patch. Wet both the plug and the insertion tool with vulcanizing fluid. Always pour directly from the can so as no to contaminate the can's contents (Fig. 48).

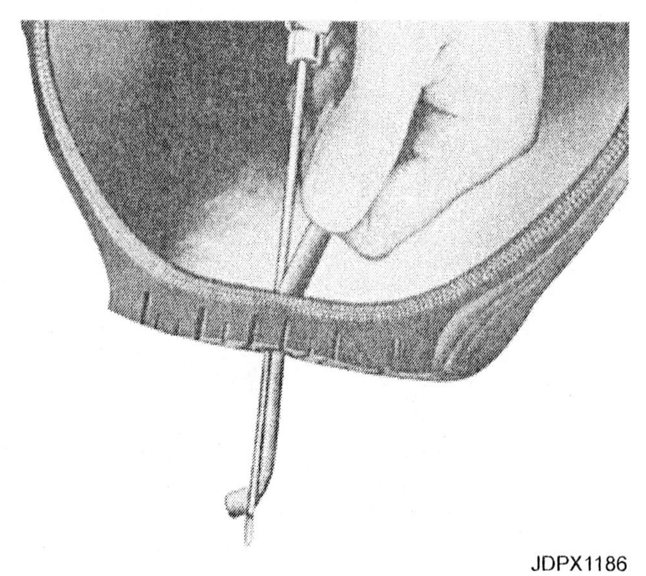
JDPX1186

Fig. 49 — Insert Plug

4. While holding and stretching the long end of the plug, insert the plug into the injury from the inside of the tire. Hold and stretch the long end of the plug as it is forced into the injury until one end extends through it (Fig. 49).

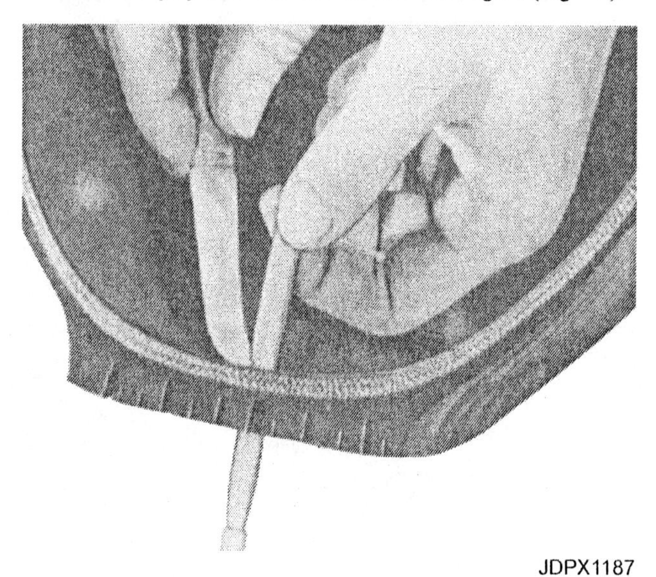
JDPX1187

Fig. 50 — Cut Off Plug

5. Remove the insertion tool and cut off the plug 1/16 inch (1.5 mm) above the surface (Fig. 50). Do not pull on the plug while cutting it. Never wash the previously prepared surface with solvent prior to the application of vulcanizing fluid.

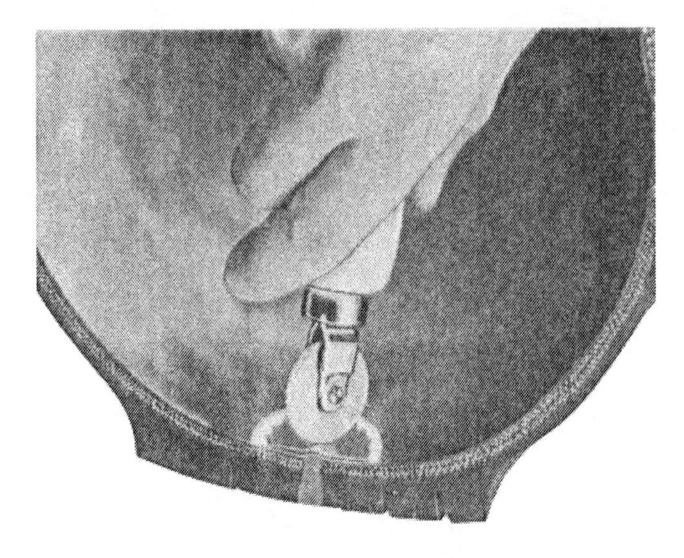

JDPX1188

Fig. 51 — Stitch Cold Patch

6. When using a cold patch, carefully remove the backing from the patch. Center the patch base on the damaged area where the vulcanizing fluid has been spread and allowed to dry. Stitch the patch down firmly with the stitching tool (Fig. 51), working form the center out.

JDPX1189

Fig. 52 — Vulcanizing Hot Patch

7. When using a hot patch, cover the buffed area with a light coat of cement, of the type specified for the patch, and allow to dry. Remove the backing from the patch. Center the patch over the injury (Fig. 52). Clamp it finger tight only, apply heat, cure, and allow to cool. Before remounting the tire, clean and deburr the rim carefully.

NOTE: Tire repairs should normally be done by a competent tire serviceman.

Repairs to tire tubes can be made with cold patches, but hot patches are always more satisfactory.

Tubeless tires can be repaired in the same manner as tube-type tires. Injuries which may lead to blowouts in tube-type tires are detected early in tubeless tires and can be repaired before extensive and costly damage results as shown in Fig. 42.

Even with the best of maintenance, cuts will still be a source of tire trouble.

Always use the correct procedure for handling and repairing tires. Closely inspect all tires at the time of inflation checks, and repair all tires having cuts that penetrate into the cord body.

Failure to make regular inspections and repairs when needed will result in further deterioration of the cord body and eventually a blowout.

JDPX1396

Fig. 53 — Cut Around Small Tire Cuts so Stones Won't Embed Themselves

Small rocks and dirt may get into cuts in the tread, and if neglected, will eventually be pounded into the cord body. To prevent this, clean out the cut with an awl or similar tool to remove stones, etc. Then, as mentioned earlier, cut away the rubber around the cut to form a cone shape extending to the bottom of the cavity (Fig. 53).

The sides of the cavity should be slanted enough to prevent stones from wedging in. Tires with cuts treated in this way may be kept in service without danger of further growth of these injuries.

Large carcass breaks over one-third the width of the tire cannot be economically repaired.

When an injury is repairable, the anticipated life of the tire must be compared to the cost of the repair work. The older the tire, the less service is received from repairs.

RECAPPING TIRES

NOTE: Tubeless tires can be recapped in the same way as tube-type tires.

Recapped tubeless tires **do not** require the installation of a tube when put back into service if the tire beads and inner liner are in good condition.

In some off-the-road operations, such as soft dirt excavation and hauling, recapping pays big dividends.

The deciding factor is the severity of the job the tire must do. Tires on free-rolling wheels of scrapers, all grader tires, farm implement tires, etc., generally are recappable.

Some industrial jobs are too tough for recapped tires: High speed and overload operations and abnormally long service at low inflation all take too much life out of the cord body to expect it to last longer than the life of more than one tread. An exception is the case of fast tread wear from steep grades and abrasive surfaces. Where these conditions exist, recapping is highly profitable.

On large industrial tires with wire in the body, recapping is advantageous. Modern recappers can recap wire and will replace the wire if necessary.

Storing and Handling Tires

Tires and tubes can deteriorate unless they are stored carefully. It is also important to handle them with care to prevent damage.

STORING AND HANDLING TUBELESS TIRES

Store and ship unmounted tires in a vertical position. Horizontal storing may compress the beads and make inflation difficult.

Do not lift tires by the beads with hooks or forks. Sharp hooks or forks may tear, cut, or snag the tubeless tire beads and result in leaks.

Remove all moisture and other foreign matter from the inside of the tire.

Tubeless tire rims perform an important function as a part of the assembly air seal. For this reason, be careful not to distort or mutilate the rim parts. Always check the O-ring groove for burrs, rust, and excess paint. Never lift the rim by the valve hold. Never drop, tumble, or roll rim parts.

Use babbit or lead hammers, not sledge hammers, when assembling rims.

Store O-ring seals carefully in a cool, dry place where they will not be injured or damaged.

Store valves in a cool, dry, clean place.

STORING TUBE-TYPE TIRES

Tires and tubes can rapidly deteriorate under certain conditions. They must be stored with extra care. Limit storage time by using tires in the same order as they are received.

Factors which cause deterioration in storage area: light, heat, airdrafts, ozone, oils, dust, dirt, and water inside the tires.

The following guidelines are recommended for tire storage.

Storing New Tires

- Store new tires indoors in a cool, dark, dry place, free from drafts. If indoor storage is impractical, tires can be stored outdoors, covered with opaque tarpaulins. Water and moisture must be kept out of tires. This can best be done by mounting them on wheels, inflating to 50 percent of operating pressure, and covering them with tarpaulins.
- Store tires away from electrical devices such as motors or switches which are an active source of ozone.
- Do not store tires in the same or adjoining rooms with gasoline and lubricants. The solids, fluids, or vapors from them are readily absorbed by rubber and cause rapid deterioration.
- Tubeless tires should not be stacked but should be stored in a vertical position, on the tread. Stacking them will deform tires, force the beads together, and produce strains in the rubber which will accelerate weather damage and make inflation difficult. New, tubeless tires are banded to prevent bead deformity. Do not remove the bands until you mount the tire.

Storing Used Tires

- Clean and carefully inspect used tires before storing. Make all necessary repairs before storing, especially if the cord fabric is exposed, as moisture will be absorbed readily.
- Store used tires the same way you store new ones.

Storing Mounted Tires

- If tires must be stored while mounted on the vehicle, block up the vehicle so weight does not rest on the tires and release the air from the tires. When the vehicle cannot be blocked, check air pressure frequently and maintain the FULL inflation.

- Vehicles on tires should be moved occasionally so the same section of the tire is not always under strain from deflection.
- Protect each tire with a cover or wrapping of canvas or similar material.
- *Never* use paint to preserve tires. If exposure is severe, consult your tire supplier for recommendation.

Storing Tubes

- New tubes should be left in their original packages. Store in a dry, cool, draft-free place.
- Used tubes should be removed from the tire, completely deflated, cleaned, folded, and stored in the same manner as new tubes.

Tire Mounting and Demounting

We will describe mounting and demounting the following four types of tires:

- Small Tires for Implements and Trucks
- Large Tires for Implements and Trucks
- Tractor and Machine Drive Wheel Tires
- Large Tires for Off-The-Road

Follow manufacturer or manufacturer association procedures to mount or demount a tire from a wheel and rim.

JDPX1190

Fig. 54 — Explosive Separation of Tire and Rim Parts Can Cause Serious Injury or Death

CAUTION: Failure to follow proper procedures when mounting a tire on a wheel or rim can produce an explosion which may result in serious injury or death. Do not attempt to mount a tire unless you have the proper equipment and expertise to perform the job.

When seating tire beads on rims or wheels, never exceed maximum inflation pressures specified by tire manufacturers for mounting tires. Inflation beyond this maximum pressure may break the bead, or even the rim, with dangerous explosive force.

Various tire sizes and designs may have different maximum inflation pressure limitations for safety. Know what these maximum pressures are before attempting to inflate a tire.

Safety Tips for Inflating and Mounting Tires

The following safety guidelines are general recommendations. Different designs of tires and rims may require different safety practices for inflating and mounting procedures.

Before working on any tire, or before removing it from the vehicle, remove the valve stem and drain all air. Insert a small wire through the valve to make sure there is no blockage.

If you don't follow proper procedures when inflating or mounting a tire, you may be seriously injured or killed.

CAUTION: Always use protective eyewear whenever using compressed air to prevent eye injuries caused by air-blown particles.

- Deflate tires on multipiece wheels before you remove the wheel if there is tire or wheel damage, or if the tire was driven with less than 80 percent air pressure.
- Unless you have the proper tools and experience, it is unsafe to attempt to inflate or mount tires. Have a tire service station repair it.
- Follow machine manufacturer's recommendations for removing tire and wheel or rim assemblies.
- Regardless of how firm the ground appears, place sound wood blocks under the jack. Don't use masonry blocks or bricks because they could break under load.
- Handle tire and wheel or rim assemblies carefully. Some assemblies are heavy and can crush your arms or legs if they slip.
- Support the machine safely. Before jacking the equipment off the ground, take steps to keep it from rolling and falling. If the machine is self-propelled, put the transmission in gear or in park position, set the parking brake and block the wheels. Place blocks or stands under the equipment immediately after it has been jacked up.
- Get help when handling large tires. Big, heavy tires can tip and fall, injuring your back, straining muscles, or breaking bones. When handling large tires, stand at the side of the tire, not in front of the tire. If you have a hoist, lift truck, or loader-equipped tractor, use it to handle large tires and liquid filled tires.
- When using bead breakers and tire tools, keep your fingers and feet clear of pinch points to avoid injury.

JDPX1399

Fig. 55 — Make Sure Tire Is Supported Before Putting Hands Between Tire and Rim

- Watch for pinch points. Whenever you are working with a partially mounted tire, block the tire securely to avoid being pinched or crushed between the tire and the rim, especially when installing or removing the tube (Fig. 55).
- Never place your fingers between the tire bead and the rim when inflating a tire. The beads usually "pop out" against the flange with crushing force.

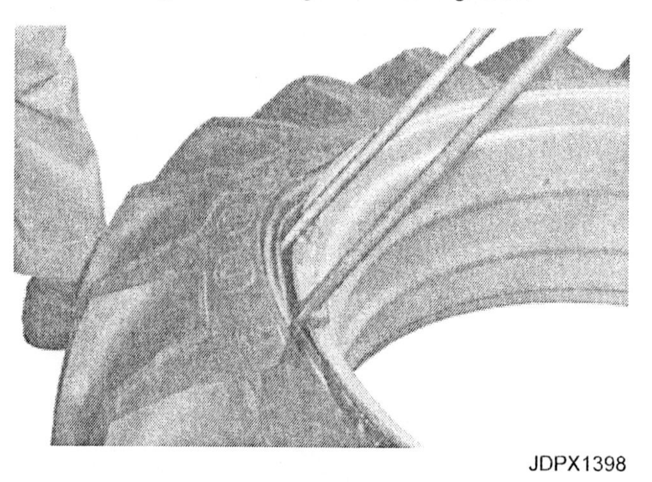

JDPX1398

Fig. 56 — Take Small Bites with Tire Irons to Avoid Hard Prying

- When using tire irons, don't try to take large sections of the tire from the rim with each bite. Instead, take smaller bites to avoid hard prying (Fig. 56). Keep your balance. Tire irons may slip, forcing you to fall.
- Keep a firm hold on tire irons or other tools. In some cases if a tool slips from your control, it could be thrown with enough force to cause serious injury.
- Inspect the inside of the tire for dirt, liquids or other foreign material and remove before installing the tube. Make sure that both the tire and tube are com-

pletely dry, as moisture may deteriorate tire cord fabric and may result in eventual failure of the tire.

- Inspect the inside of the tire for loose cords, cuts, penetrating objects or other damage and repair or replace the tire as necessary.
- Be sure the flange area (especially the bead seat area) of the rim is clean and smooth. Remove rust and corrosion with a wire brush and repaint the rim or coat it with a rust inhibitor to prevent corrosion.
- Never weld or heat a wheel and tire assembly. The heat can cause an increase in the tire inflation pressure, resulting in a tire explosion. Welding can also structurally weaken or deform the wheel.
- Never, under any circumstances, attempt to rework, weld, braze or heat any rim components that are cracked, broken or damaged.
- Never mount a tire on a rim where any parts show cracks, damage, or have been repaired by welding or brazing.
- Do not mix parts of one type rim with those of another. The parts may appear to fit, but when the tire is inflated they can fly apart with explosive force.
- Be sure all components are properly seated before inflating the tire. DO NOT hammer to seat split rings or other components while the tire is fully or partially inflated.

JDPX1397

Fig. 57 — Apply Rubber Lubricant to Help Slip Tire Bead Over the Rim

- Before mounting a tire, lubricate the tire bead, when recommended, with a solution of vegetable oil soap in water or with an equivalent approved tire mounting lubricant (Fig. 57). Never use petroleum-base or silicone lubricants. The use of a lubricant will allow easier seating of the bead and reduce the pressure necessary to seat it.

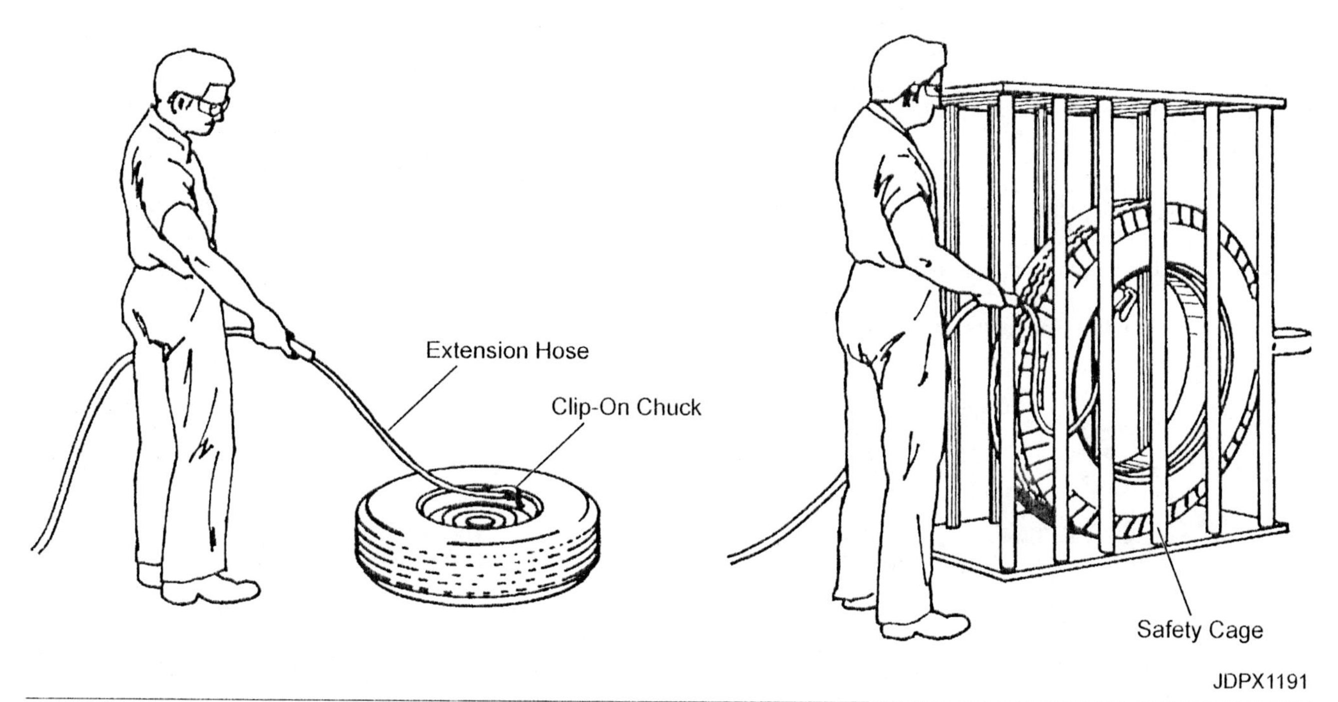

Fig. 58 — Use a Clip-On Chuck Hose Extension when Inflating Tires

- ALWAYS use a safety cage or equivalent restraining device when inflating the tire to seat the beads and/or inflating the tire to normal operating inflation pressure (Fig. 58).
- When inflating tires, use a clip-on chuck and an extension hose long enough to allow you to stand to one side and NOT in front of or over the tire assembly (Fig. 58). The inflation hose must be equipped with an in-line air valve and pressure gauge.
- If either bead has not seated by the time the inflation pressure reaches the tire manufacturer's specified maximum pressure, do not increase the inflation pressure to seat the bead. Deflate the tire, reposition the tire on the rim, lubricate the beads and rim with an approved rubber lubricant, and reinflate the tire. Inspect both sides of the tire to be sure the beads are evenly seated. If not, deflate the tire, unseat the beads and repeat the mounting procedure.
- Never, under any circumstances, introduce a flammable substance into a tire before, during or after mounting. Igniting this substance in an effort to seat the tire beads is extremely unsafe and could result in an explosion of the tire with a force sufficient to cause serious personal injury or death. This practice could also result in undetected damage to the tire or rim that could result in failure of the tire in operation.

The following procedures for mounting and demounting tires are intended only to familiarize you with proper methods, not to make you an expert. Do not attempt to demount or mount tires without training by an experienced person who is familiar with the proper methods and tools.

Changing Small Tires for Implements and Trucks

TOOLS

Unless you have a tire mounting machine, you will need a proper set of tire tools. It is very easy to damage the tubeless tire beads or the inner tube when changing tires unless you have the proper tools.

TIRE BEAD LUBRICANTS

Vegetable oil and animal oil soap solutions may be used as a lubricant to mount tires. The lubricant should also contain a rust inhibitor if it is water-based. When dry, the lubricant must not remain slippery. To minimize moisture in the tire chamber, avoid excessive application of lubricant.

Silicone, petroleum or solvent based lubricants MUST NOT be used. These substances may cause the tire to slip on the rim. They may have a harmful effect on the tire, tube, flap or rim. They may create explosive mixtures of air and volatiles in the tire which may result in serious injury or death.

DEMOUNTING SMALL TIRES FOR IMPLEMENTS AND TRUCKS

Contact tire, wheel, and rim manufacturers and associations for specific information.

With a Tire Mounting Machine

1. Remove valve core and completely deflate tire.

2. Apply lubricant at top of rim flange.

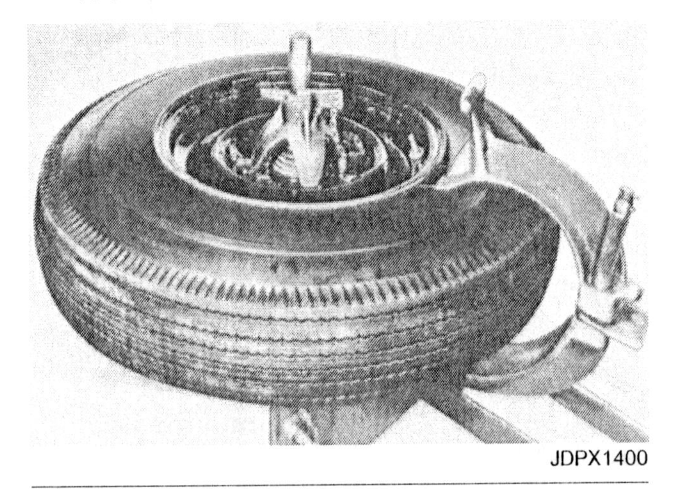

JDPX1400

Fig. 59 — Loosen Tire Beads Using a Bead Breaker Tool

3. Loosen both tire beads from rim flanges with bead breaker tool (Fig. 59).

CAUTION: Do not use a hammer or tire iron to loosen beads. Do not damage any portion of beads.

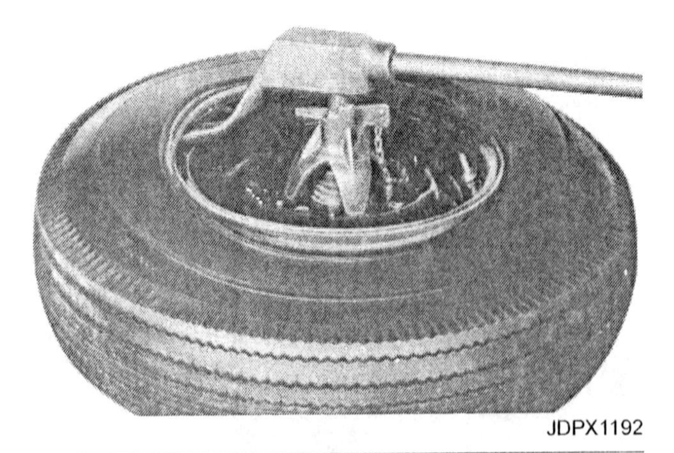

JDPX1192

Fig. 60 — Removal of First Bead

4. Press side of tire into rim well and remove first bead in regular manner (Fig. 60). (On tube-type tires, remove the inner tube — carefully.)

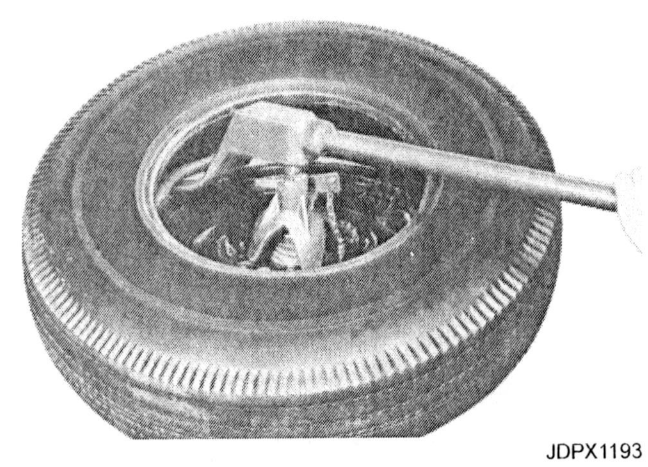

JDPX1193

Fig. 61 — Removal of Lower Bead

5. Remove the lower bead (Fig. 61). Then remove the tire from the rim.

Without a Tire Mounting Machine

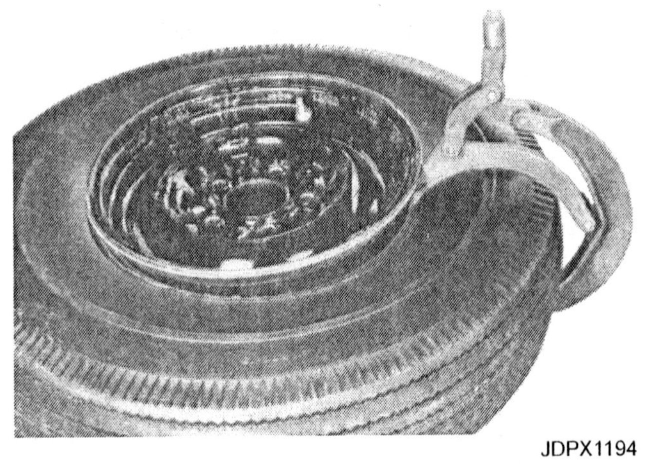

JDPX1194

Fig. 62 — Loosen Tire Beads Using a Commercial-type Bead Breaker Tool

1. Remove valve cord and completely deflate tire.

2. Apply lubricant at top of rim flange.

3. Force bead away from bead seat using foot pressure. If tire bead cannot be forced away by foot pressure, use any commercial-type bead breaker tool (Fig. 62). Do not use hammer or tire irons to loosen beads. Do not damage the beads.

Fig. 63 — Removal of Outside Bead

4. Use clean, smooth tire irons carefully to remove outside bead. Start removal of outside bead at valve, taking small bites with irons around rim (Fig. 63). Do not damage the bead area. (On tube-type tires, remove the inner tube — carefully.)

Fig. 64 — Pry Rim Upward through Inside Bead

5. Turn tire over and use two irons — one between rim flange and tire bead to pry rim upward and the other iron between bead seat and tire bead to pry outward (Fig. 64). Start removal at valve to eliminate any possibility of bead catching on valve base.

PREPARING RIM BEFORE MOUNTING (TUBELESS TIRES)

Contact the wheel manufacturer or association for specific information.

Because the rim must be leakproof in a tubeless tire, the following steps must be taken **before** the tire is mounted to be sure the rim will not leak.

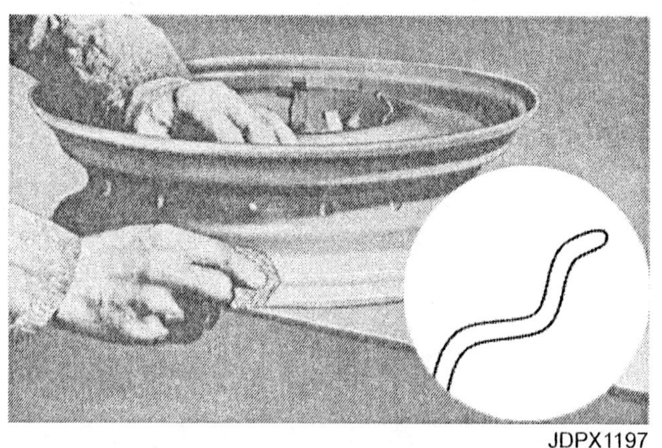

Fig. 65 — Cleaning of Rim Flange and Bead Seat

1. Rim flange and bead seats must be clean. Remove all foreign material with emery cloth or coarse steel wool. Rust can be removed with a wire brush (Fig. 65). (Circle shows area which must be clean.)

Fig. 66 — Straighten Any Visible Dents in the Rim Flange Area

2. Straighten any visible dents in the rim flange area with a hammer (Fig. 66). Smooth out any hammer dents with a file.

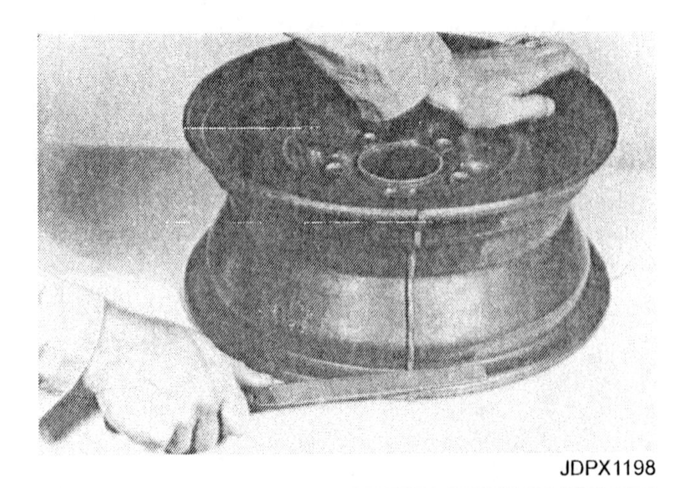

JDPX1198

Fig. 67 — Inspect for Porous Butt-weld

3. Inspect for porous butt-weld. Grooves or high spots (in bead area only) must be filed flat and smooth or tire assembly may lose air (Fig. 67).

MOUNTING SMALL TIRES FOR IMPLEMENTS AND TRUCKS

Contact tire, wheel, and rim manufacturers and associations for specific information.

In every case, be sure to follow the proper procedure for preparation of the tubeless rim before mounting the tire. Use a thin vegetable oil soap solution or other approved rubber lubricant for lubricating beads and rims. Any metal tools contacting the tire must be clean and smooth to prevent bead damage (tubeless tires).

First, in tubeless tires, install the valve in the rim. Be sure that both inside and outside of rim area around valve hole are clean and free of any burrs. Use steel wool or file. When installing rubber snap-in valve, lubricate valve before installing. Use valve inserting tool to be sure valve base is firmly seated against inside of rim surface. Be careful to avoid damage to threads or other parts of valve.

With a Tire Mounting Machine

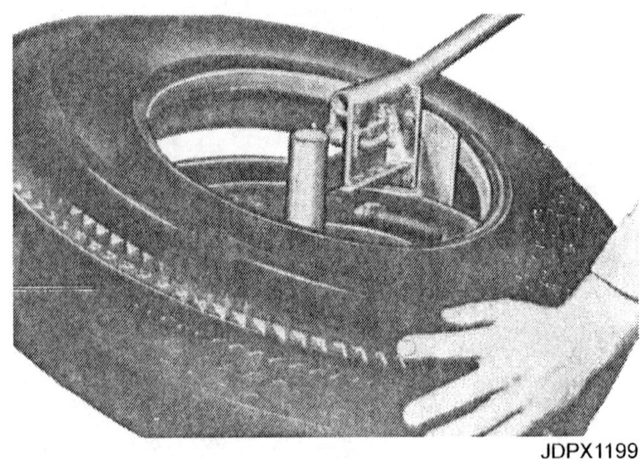

JDPX1199

Fig. 68 — Mounting of the First Bead with a Mounting Machine

1. After preparing tire and rim and installing valve, lubricate base of tire beads and the rim. Mount the first bead as shown in Fig. 68. If tire contains a tube, install the tube, making sure the valve is properly located in the valve hole. Inflate the tube just enough to almost round it out.

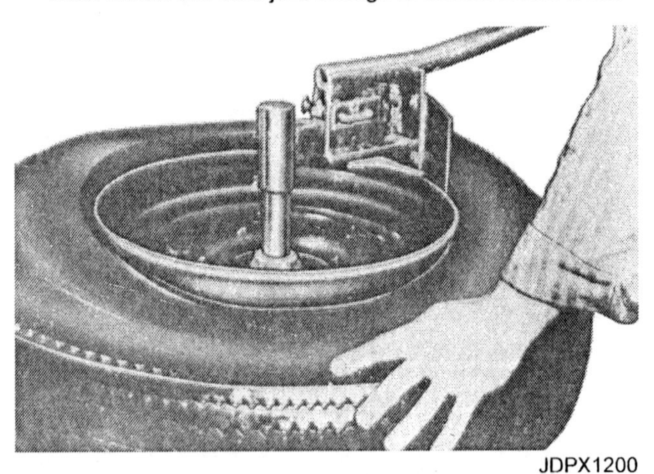

JDPX1200

Fig. 69 — Mounting of Second Bead with a Mounting Machine

2. Mount the second bead in the same manner as first (Fig. 69). Apply the portion of the bead nearest the valve last to eliminate any possibility of the bead catching on the valve base. Line up balance mark (if any) on the tire sidewall with the valve in the wheel.

Without a Mounting Machine

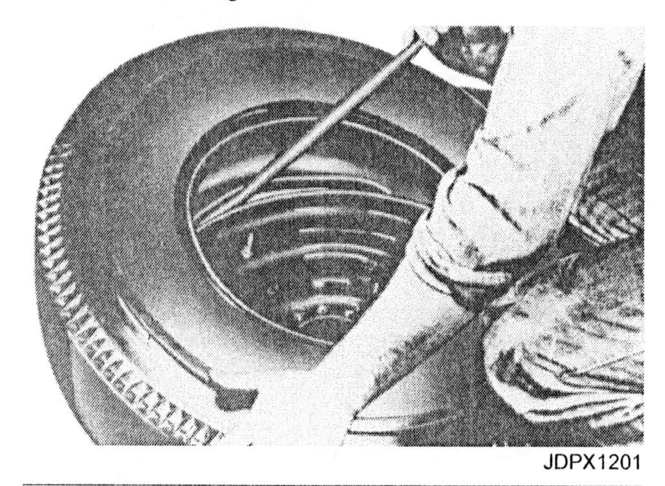

JDPX1201

Fig. 70 — Mounting of First Bead Without a Mounting Machine

1. Lubricate tire beads and rim. Mount first bead by taking small "bites" around the rim with smooth, clean tire irons. Be careful not to damage beads. If tire has a tube, mount as instructed above.

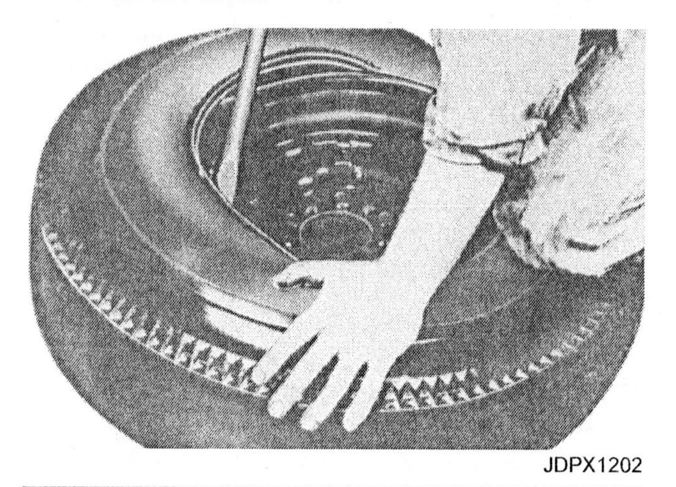

JDPX1202

Fig. 71 — Mounting of Second Bead Without a Mounting Machine

2. Mount the second bead in the same manner as first. Apply portion of bead nearest valve last. Line up balance mark (if any) on the tire sidewall with valve in wheel. Do not use a hammer.

MOUNTING INNER TUBES ON TUBE-TYPE TIRES AND INFLATING

1. While mounting, insert tube in tire and inflate until nearly rounded out.

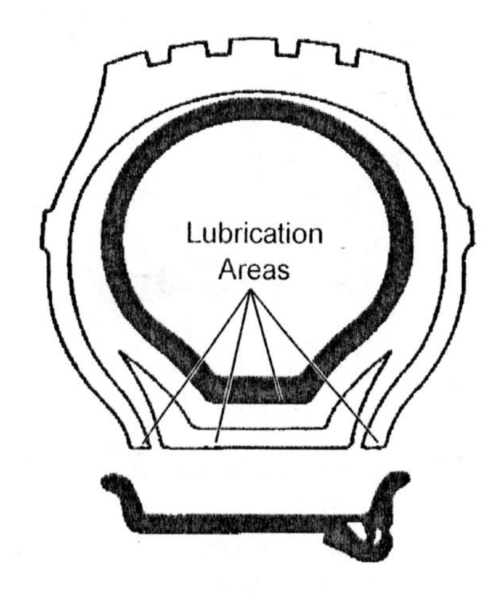

JDPX1209

Fig. 72 — Lubrication Points

2. Using a brush or cloth swab, apply a solution of neutral vegetable oil soap in the areas indicated in Fig. 72, including the flap. Do not allow the soap solution to run down into the tire.

3. Mount the tire on the rim. Center the valve, pull it through the hole in the rim, and hold it firmly against the rim. Hold the valve in this position while inflating the tire until the beads are properly seated (do not exceed manufacturer's recommended pressure).

4. Completely deflate by removing the valve core or using a deflating tool. *This is extremely important to prevent tube buckling.*

5. Reinflate to recommended pressure.

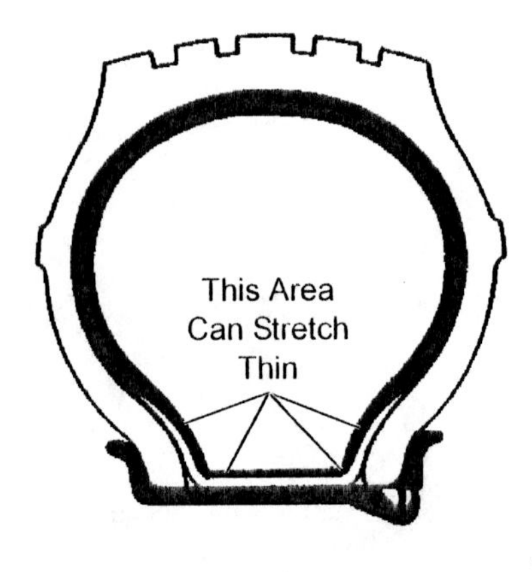

Fig. 73 — Stretched Tube and Flap

IMPORTANT: Unless the tube and flap are lubricated, they will be stretched thin in the tire bead and rim areas (Fig. 73), causing premature failure.

INFLATING TUBELESS TIRES AFTER MOUNTING

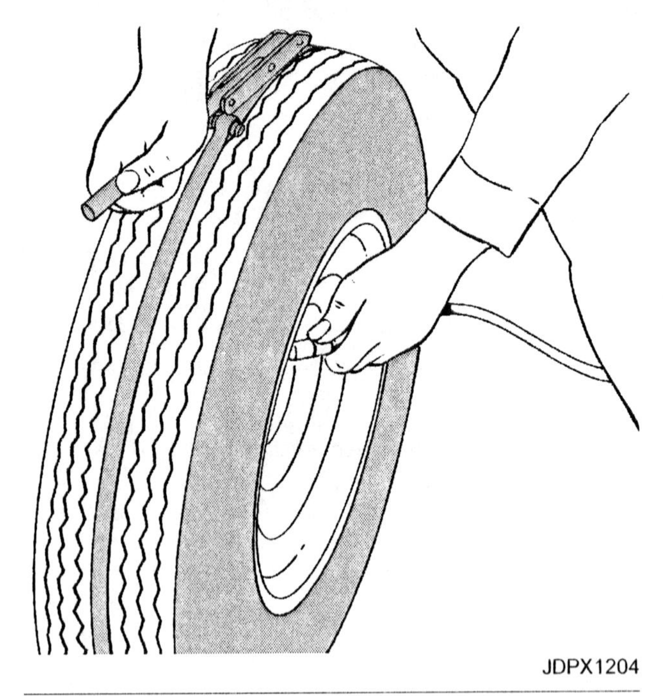

Fig. 74 — Using Tire Bead Expander while Inflating Tubeless Tire — DO NOT STAND OVER TIRE

1. After the tire is mounted on the rim, start inflating it in the usual manner (Fig. 74).

CAUTION: Do not stand over the tire while inflating it.

If the tire beads do not contact both the rim seats tight enough to retain air, spread the beads by constricting the tread centerline of the tire. This is usually done using a mounting band (bead expander) as shown in Fig. 74.

Another method of seating tubeless tire beads is the use of well lubricated, heavy rubber rings positioned between the tire beads and rim. The rings form a seal between the tire beads and rim, allowing the air pressure to move the beads out and seat properly.

2. Inflate the tire just enough to move the beads out to contact the bead seats on the rim. Then, remove the bead expander or rubber rings. Increase the air pressure as needed to fully seat the tire beads on the rim.

3. Check for leaks, then adjust the inflation pressure to the recommended operating pressure.

4. On safety or hump-type rims, make sure the tire beads have snapped over the hump on the rim and are fully seated.

CAUTION: Failure to follow proper procedures when mounting a tire on a wheel or rim can produce an explosion which may result in serious injury or death. Do not attempt to mount a tire unless you have the proper equipment and expertise to perform the job.

When seating tire beads on rims or wheels, never exceed maximum inflation pressures specified by tire manufacturers for mounting tires. Inflation beyond this maximum pressure may break the bead, or even the rim, with dangerous explosive force.

Changing Large Tires for Implements and Trucks

DEMOUNTING LARGE TIRES FOR IMPLEMENTS AND TRUCKS

Contact tire, wheel, and rim manufacturers or association for specific information.

Demounting large tubeless truck or implement tires is similar to what we describe for smaller tires.

1. Remove valve core and completely deflate.

2. Lay assembly on floor with narrow ledge on bottom.

JDPX1205

Fig. 75 — Use a Bead Unseating Tool to Unseat Tire Bead

3. Drive a bead unseating tool between tire bead and rim flange, being careful not to damage tire bead area. After bead has been released completely around tire, turn tire and rim over and repeat bead unseating procedure with the other bead.

JDPX1206

Fig. 76 — Lubricate Rim Flange and Tire Bead

4. With narrow ledge of rim well on top, thoroughly lubricate rim flange and tire bead with a thin solution of vegetable oil soap in water or equivalent rubber lubricant recommended for this requirement. (Never use petroleum-base lubricants or silicones.)

JDPX1207

Fig. 77 — Pry Bead Over Rim Flange Using 18" Long Tire Irons

5. Force part of bead across rim from valve into well. Starting at valve, pry bead over rim flange using two 18" long tire irons. Continue by taking short bites to avoid damage to bead until top bead is completely over the rim flange.

JDPX1208

Fig. 78 — Pull Tube Out of Tire Casing

6. Bring assembly to upright position and pull tube out of the tire casing. When only tube requires repair or replacement, thoroughly inspect inside of tire casing for foreign material or damage and make sure both tube and inside of casing are dry before reinserting tube.

JDPX1209

Fig. 79 — Work Rim Slowly Out of Tire Using Tire Irons

7. To completely remove tire from rim, turn assembly over and lubricate second tire bead and rim flange. Be sure on side of bead still on the rim is in the rim well and insert tire irons under opposite side of bead. Work rim slowly out of tire by taking small bites alternately using both tire irons.

MOUNTING LARGE TIRES FOR IMPLEMENTS AND TRUCKS

Following the steps outlined in section titled "Preparing Rim Before Mounting (Tubeless Tires)" to prepare the rim for mounting.

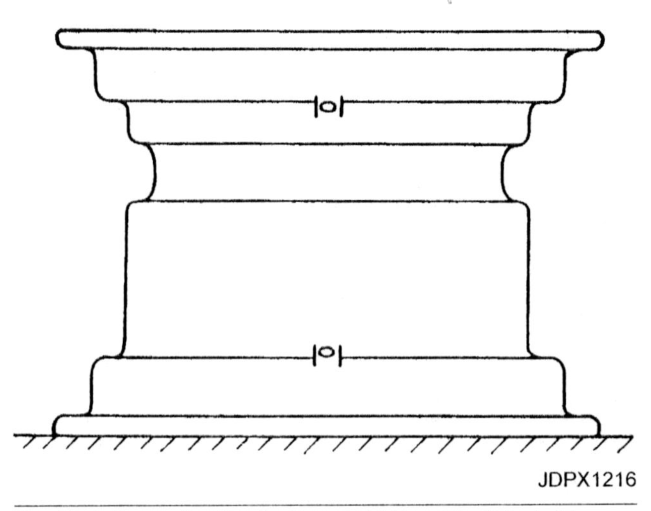

JDPX1216

Fig. 80 — Proper Position for Deep Well-Type Rim Prior to Tire Changing

IMPORTANT: **On deep well rims, the deep well must be positioned up as shown in Fig. 80. Start the tire bead on the rim nearest the deep well. If attempts are made to mount the tire over the long side of the rim first, severe damage to the tire bead could result.**

JDPX1210

Fig. 81 — Lubricate Bottom Tire Bead and Top Rim Flange

1. Lay rim on floor with narrow ledge on top. Lubricate bottom tire bead and top rim flange with a thin solution of vegetable oil soap in water or equivalent rubber lubricant recommended for this requirement. (Never use petroleum-base or silicone lubricants.)

JDPX1213

Fig. 82 — Use Tire Irons to Work First Bead Over Rim Flange

2. Push bottom bead over rim flange as far as possible. Use 18" tire irons to work first tire bead completely over rim flange, taking small bites and being careful not to damage bead. Use 36" irons on large rears.

NOTE: On deep well rims, start bead on rim nearest deep well (Fig. 80).

JDPX1211

Fig. 83 — Use Valve Retrieving Tool to Position Valve in Hole

3. Partially inflate tube and insert in tire casing with valve located near valve hole in rim. Attach valve retrieval tool to valve and thread tool through the valve hole. (Inserting the tube and attaching the tool may be facilitated by placing a block under the tire.)

JDPX1214

Fig. 84 — Work Top Bead Over the Rim Flange

4. Starting opposite the valve, use tire irons to lever top bead over the rim flange and down into rim well. Be careful to avoid pinching tube with tire irons. Locking pliers may be used to "keep your place." When bead is well started, lubricate remaining unmounted portion of tire bead and rim flange. Taking small bites, spoon tire bead over rim flange until final section drops over at valve.

JDPX1212

Fig. 85 — Lubricate Beads and Bead Seats on Both Sides of Tire

5. Thoroughly lubricate tire beads and rim bead seats on both sides of tire.

Fig. 86 — Inflate Tire to Seat Bead

6. Center tire on rim and inflate to fully seat bead using an extension hose with clip-on chuck and gauge to permit operator to stand clear of tire. **(Do not exceed 35 psi.)** Then remove valve core and completely deflate. Re-insert valve core and reinflate tire to recommended pressure.

NOTE: If either bead should fail to seat at 35 pounds inflation, the tube may be pinched between tire bead and rim or something else is interfering with proper mounting. Do not increase inflation pressure to seat beads, but remove valve core and completely deflate. Break both beads loose from rim, re-lubricate both tire beads and rim bead seat areas. Reinstall core and repeat inflation procedures.

Mounting Large Tires onto Disk Wheels

With disk wheels, the mounting procedure is the same. On certain larger disk wheels, the dish is so deep that it is difficult to insert the tire tool in the normal way. With the first bead in place, clamp on the locking pliers in the usual manner. Then, to get around the deep dish of the wheel, insert the spoon end of the tool at an angle between the dish and the flange. Swing the tool toward the center of the dish and work the bead over the flange as usual.

If the tire uses a tube and flap, after the first bead has been placed on the rim, install the tube and flap into the tire. Inflate the tube just enough to round it out and, with the valve stem in the valve hole in the rim, install the valve stem retaining nut and draw it up snug. Install the second bead onto the rim as instructed in "Mounting Large Tires For Implements and Trucks".

NOTE: Most larger tubeless tires cannot be mounted on conventional rims.

INFLATING TUBELESS TIRES AFTER MOUNTING

CAUTION: Always use protective eyewear whenever using compressed air to prevent eye injuries caused by air-blown particles.

1. Check the valve to be sure the retaining nut at the base is tight. Also check the valve core to make sure it is seated tight in the valve.

2. Center the tire on the rim.

3. Place the assembly in a safety cage or other approved restraining device, then inflate using an extension hose with a clip-on chuck and air pressure regulator valve. Inflate to fully seat the beads.

CAUTION: Do not stand over the tire while inflating.

CAUTION: Do not exceed the maximum pressure specified by the tire manufacturer for mounting tires.

INFLATING TUBE-TYPE TIRES AFTER MOUNTING

1. Place the tire and rim assembly in a safety cage or other approved restraining device, then inflate the tire using an extension hose with a clip-on chuck and air pressure regulator valve.

2. Inflate until the beads are properly seated, then remove the valve core and completely deflate the tire. This removes the buckles and uneven stress on the tube and flap.

3. Then, install the valve core and reinflate to the proper operating pressure.

If the tube and flap are not properly lubricated and mounted, they will be stretched thin in the tire and bead area (Fig. 73). This will cause premature failure.

Changing Agricultural Tractor Tires

Guidelines for mounting and demounting agricultural tractor tires are shown in this section. It is important that specific tire, wheel, and rim information is used to mount and demount these tires. See the manufacturers or manufacturer associations for this information. Proper tools, training, and procedures must be used. Demounting and mounting a tire improperly is hazardous.

CHAINS, HOISTS, AND SUPPORTS

Large tires, rims, and wheels are extremely heavy and awkward to handle. They must be properly supported with chains, hoists, or other supports. Failure to do this can result in serious injury.

TIRE, RIM AND WHEEL ASSOCIATIONS

CAUTION: Every tire and rim or wheel must be handled in a special way. Always use the tire and rim and wheel manufacturer's procedures when you demount and mount a tire.

Information also can be obtained from the following associations:

Rubber Manufacturer's Association
1400 K Street, N.W.
Washington, D.C. 20005

National Wheel and Rim Association
4836 Victor Street
Jacksonville, FL 32207

Further safety information can be obtained from:

U.S. Department of Transportation
National Highway Traffic Safety Administration
400 Seventh St., S.W.
Washington, D.C. 20590

DEMOUNTING AGRICULTURAL TRACTION TIRES (ON MACHINE)

Contact tire, wheel, and rim manufacturers and associations for specific information.

1. With the vehicle jacked up on the side which is to have the tire removed, and after fluid has been removed, remove the rim nut and core housing to completely deflate. Push valve through valve hole.

JDPX1217

Fig. 87 — Unseat Beads Using a Bead Breaking Tool and a Heavy Hammer

2. Using a bead breaking tool and a heavy hammer, drive the tool between the tire bead and the rim flange, being careful not to damage bead area (Fig. 87). Beads should be unseated on both sides of rim.

JDPX1218

Fig. 88 — Use of a Bead Unseating Tool

As an alternative to the procedure above, a bead unseating tool and instructions are available from Iowa Mold Tooling Company (Fig. 88).

JDPX1219

Fig. 89 — Lubricate the Rim Flange, Tire Bead and Base of Tube

3. Thoroughly lubricate rim flange, tire bead and base of tube with a thin solution of vegetable oil soap in water or equivalent rubber lubricant recommended for this requirement (Fig. 89). (Never use petroleum-base solution or silicones.)

JDPX1220

Fig. 90 — Pry First Section of Bead Over Rim Flange

4. Lock wheels by putting vehicle into gear with valve at top. Force outside bead at bottom into rim well. Insert long tire irons under bead at the top and pry bead over rim flange (Fig. 90). Take short bites to avoid extremely hard prying and possible damage to tire bead.

JDPX1221

Fig. 91 — Pry Next Section of Bead Over Rim Flange

5. After first section of bead is over rim flange, use one iron to hold that section over flange. Use the other tire iron to pry sections until the entire bead is over flange (Fig. 91). Do not attempt to pry too large a section over the rim flange at one time.

CAUTION: Never release grip on either iron as they may tend to spring back.

JDPX1222

Fig. 92 — Pull Tube Out of Casing

6. Pull tube out of casing (Fig. 92). When only tube requires repair or replacement, you don't need to remove tire completely from rim.

JDPX1223

Fig. 93 — Thoroughly Inspect Inside of Casing

7. Thoroughly inspect inside of casing for foreign material or damage and remove moisture from inside of tire (Fig. 93).

JDPX1224

Fig. 94 — Prying of Inside Bead Over Rim Flange for Complete removal of Tire

8. To remove tire completely from wheel, insert tire irons under the rest of inside bead over rim flange (Fig. 94). When starting this operation, be sure that the bead area on opposite side of tire is down in rim well.

STORAGE RECOMMENDATIONS

For longest life, tires should be stored unmounted in a cool, dark location — on tread or stacked evenly on the side, but without other material or equipment on top. Implements should not be stored with weight on the tires.

MOUNTING AGRICULTURAL TRACTION TIRES (ON MACHINE)

Contact tire, wheel, and rim manufacturers and associations for specific information.

CAUTION: Failure to follow proper procedures when mounting a tire on a wheel or rim can produce an explosion which may result in serious injury or death. Do not attempt to mount a tire unless you have the proper equipment and expertise to perform the job. When seating tire beads on rims or wheels, never exceed maximum inflation pressures specified by tire manufacturers for mounting tires. Inflation beyond this maximum pressure may break the bead, or even the rim, with dangerous explosive force.

Before mounting a tire on a used rim, be sure the flange area and particularly the bead seat area are clean and smooth. Remove any buildup of rust, corrosion or old rubber with chisel or wire brush. Bent, cracked or otherwise damaged rims should be repaired or replaced.

Thoroughly inspect inside of casing for foreign material or damage.

Lubricate both beads with a thin solution of vegetable oil soap in water or equivalent rubber lubricant recommended for this requirement. (Never use petroleum-base solutions or silicones.)

NOTE: Some steps of the following procedure may require two men on larger size tires.

JDPX1225

Fig. 95 — Force Inner Bead Over Flange

1. Place inner tire bead over flange at top. Be sure bead is not on bead seat but is guided into well as tire irons are used to work down either side, forcing rest of inner bead over flange (Fig. 95).

2. With first bead on rim, pull tire toward outside of rim as far as possible to make room for tube. Be sure inside of tire is completely dry.

JDPX1226

Fig. 96 — Place Valve in Valve Hole and Screw Rim Nut on Stem

3. Before inserting tube in tire, be sure that valve hole is at bottom of wheel. Align stem with valve hole and place tube in tire starting at the bottom. Place valve in valve hole and screw rim nut on stem to hold it in place (Fig. 96). Be sure that tube is well inside rim before proceeding to next step.

JDPX1227

Fig. 97 — Lubricate Partially Inflated Tube to Prevent Localized Stretching

Partially inflate the tube. Lubricate the base area that contacts rim to prevent localized stretching (Fig. 97).

JDPX1228

Fig. 98 — Lift Outer Bead Over Rim Flange and Into Rim Well

Starting at top, use tire irons to lift outer bead up and over rim flange and down into rim well (Fig. 98). Be careful not to pinch tube in the operation.

JDPX1229

Fig. 99 — Pry Remainder of Bead Over Flange with Tire Iron

After getting first section of outer bead into rim well, place one hand against that section to hold it, and then pry remainder of bead over flange with tire iron in other hand (Fig. 99). Do not attempt to pry large sections of bead over flange at one time.

JDPX1230

Fig. 100 — Inflate Tire to Seat Beads on Rim at Bottom

CAUTION: Centering the tire is extremely important to prevent broken beads. Remote control inflation equipment should be used. Never stand in front of assembly.

4. With valve stem at bottom, lower the jack until tire is centered on rim. Inflate sufficiently to seat beads on rim at bottom (Fig. 100).

JDPX1231

Fig. 101 — Inflate Tire to Fully Seat Beads

5. Raise tractor; rotate tire so valve is at top. Inflate to tire manufacturer's specifications to fully seat beads (Fig. 101).

6. Then, remove valve core housing and completely deflate.

7. Reinsert valve core housing and reinflate tire to recommended pressure.

NOTE: *If either bead fails to seat at recommended inflation, the tube may be pinched between tire bead and rim, or something else is interfering with proper mounting. Do not increase inflation pressure to seat beads, but remove valve core housing and completely deflate tube. Unseat both beads from the rim; relubricate both tire beads and rim bead seat areas. Reinstall valve core housing, and reinflate tube to tire manufacturer's specifications. Repeat process until both beads are properly seated.*

8. Reduce to operating pressure before putting tire in service.

DEMOUNTING AGRICULTURAL TRACTION TIRES (OFF MACHINE)

Contact tire, wheel, and rim manufacturers and associations for specific information.

To dismount tire:

1. Completely deflate tube by removing core housing and remove rim nut.

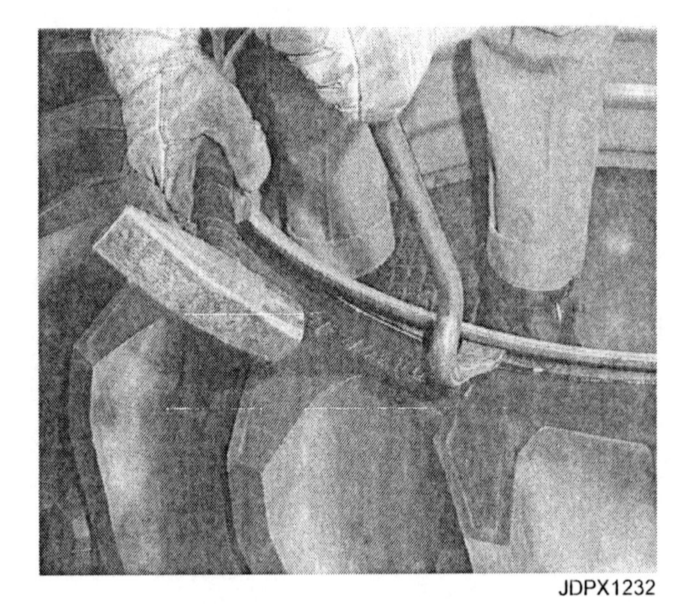

JDPX1232

Fig. 102 — Unseat Beads Using a Bead Breaking Tool and a Heavy Hammer

2. Using a "bead breaking" tool and a heavy hammer, drive the tool between tire bead and rim flange, being careful not to damage bead area (Fig. 102). Release bead completely around tire.

3. Turn tire and rim over and repeat above procedure with the second bead.

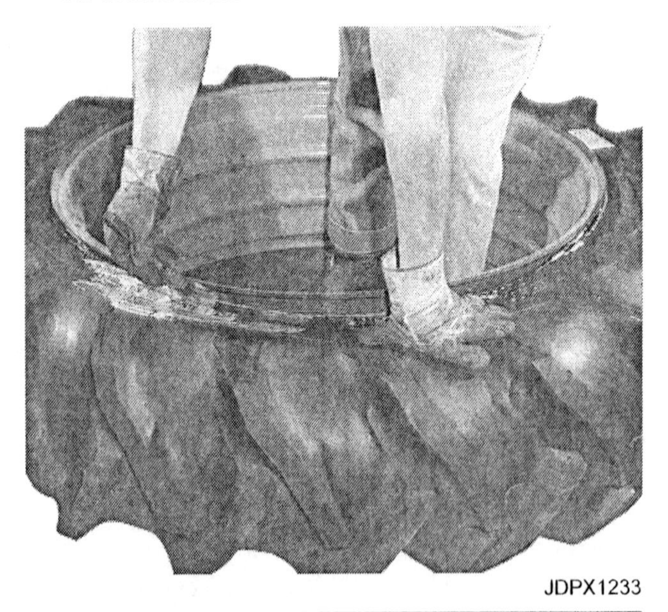

JDPX1233

Fig. 103 — Lubricate the Rim Flange, Tire Bead and Base of Tube

4. Thoroughly lubricate rim flange, tire bead and base of tube with a thin solution of vegetable oil soap in water or equivalent rubber lubricant recommended for this requirement (Fig. 103). (Never use petroleum-base solution or silicones.)

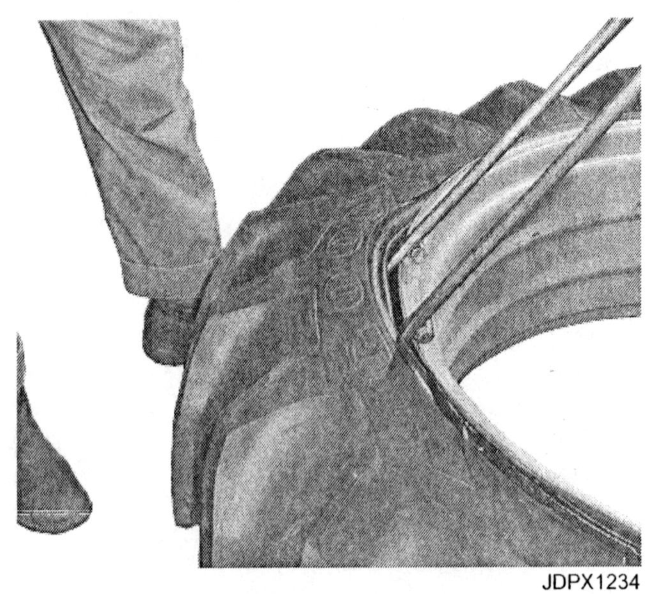

JDPX1234

Fig. 104 — Pry Top Bead Over Rim Flange

5. With part of top bead forced into rim well, pry opposite side of bead over rim flange using two long tire irons (Fig. 104). Continue until top bead is completely over the rim flange.

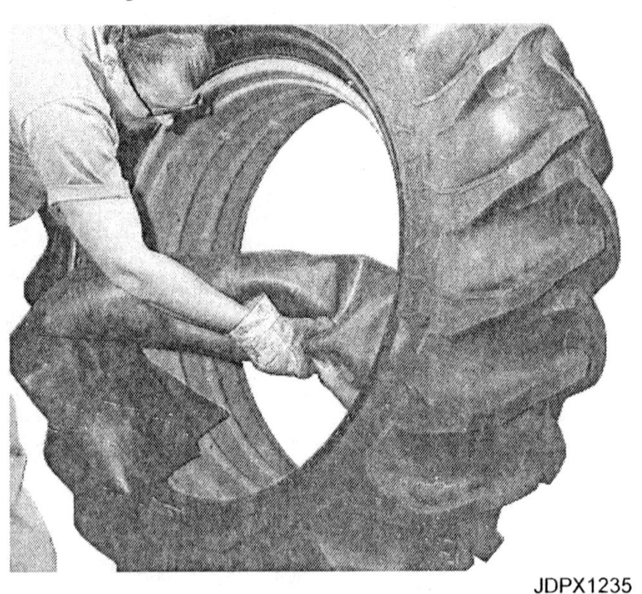

JDPX1235

Fig. 105 — Pull Tube Out of Casing

6. With weight of tire braced against a solid support, pull tube out of tire casing (Fig. 105). When only tube requires repair or replacement, thoroughly inspect inside of tire casing for foreign material or damage and make sure inside of casing is dry before reinserting tube.

JDPX1236

Fig. 106 — Prying of Inside Bead Over Rim Flange for Complete Removal of Tire

7. To completely remove tire from rim, with weight of assembly supported, be sure one side of bottom bead is in rim well and insert tire irons under opposite side of bead. With smaller size tires, work bottom bead over rim flange by taking small bites with two tire irons.

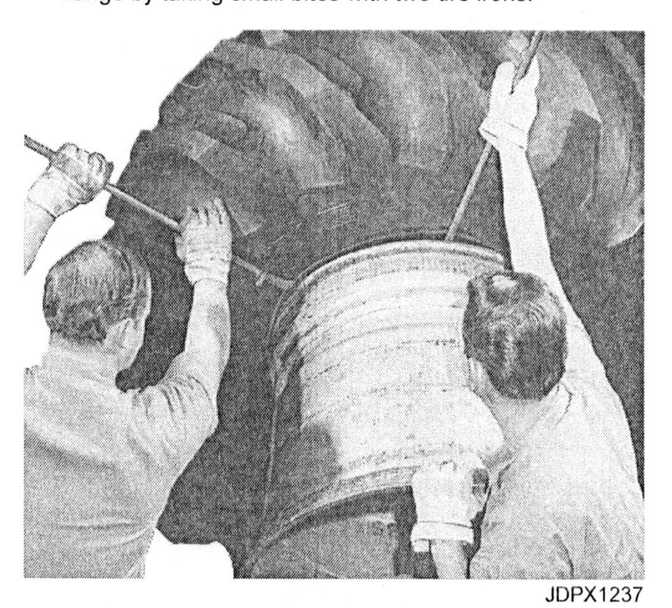

JDPX1237

Fig. 107 — Complete Removal of Large Tire from Rim

Stand large section heavy tires on tread. With weight of assembly supported, and one man holding rim, the second man can work second bead over rim flange until rim drops out.

STORAGE RECOMMENDATIONS

For longest life, tires should be stored unmounted in a cool, dark location — on tread or stacked evenly on the side, but without other material or equipment on top. Implements should not be stored with weight on the tires.

MOUNTING AGRICULTURAL TRACTION TIRES (OFF MACHINE)

Contact tire, wheel, and rim manufacturers and associations for specific information.

CAUTION: Failure to follow proper procedures when mounting a tire on a wheel or rim can produce an explosion which may result in serious injury or death. Do not attempt to mount a tire unless you have the proper equipment and expertise to perform the job. When seating tire beads on rims or wheels, never exceed maximum inflation pressures specified by tire manufacturers for mounting tires. Inflation beyond this maximum pressure may break the bead, or even the rim, with dangerous explosive force.

Before mounting a new tire on a used rim be sure the flange area and particularly the bead seat area are clean and smooth. Remove any buildup of rust, corrosion, or old rubber with chisel or wire brush. Bent, cracked or otherwise damaged rims should be repaired or replaced.

Lay the rim on the floor with the valve hole on the top side. In the case of a deep well rim, the deep well must always be on top.

1. Inflate the tube until it is rounded out and insert it in the tire with the valve on the top side before starting to mount tire on rim.

2. Lubricate bottom tire bead and top rim flange with a thin solution of vegetable oil soap in water or equivalent rubber lubricant recommended for this requirement (never use petroleum-base solutions or silicones).

JDPX1238

Fig. 108 — With Tire Cocked on Rim, Place Valve in Valve Hole and Screw Rim Nut on Stem

3. Push bottom bead over rim flange as far as possible. With the tire cocked on rim, place valve in valve hole and screw rim nut on stem to hold it in place (Fig. 108).

JDPX1239

Fig. 109 — Work First Tire Bead Completely Over Rim Flange

4. Using long tire irons, work first tire bead completely over rim flange (Fig. 109).

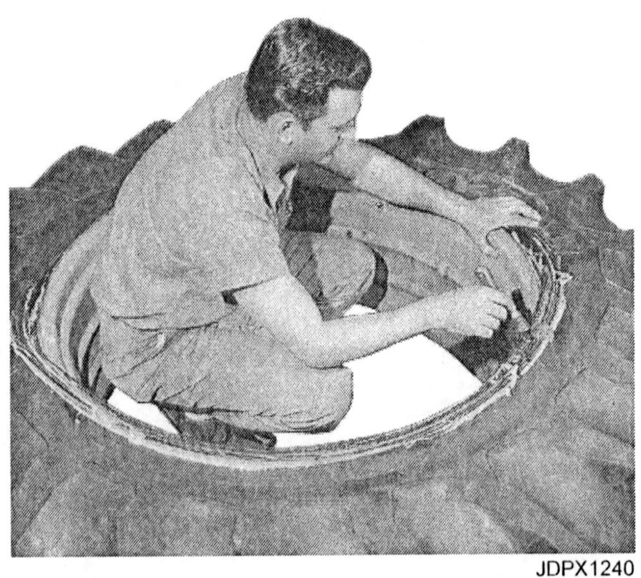

JDPX1240

Fig. 110 — Lift Top Bead Over Rim Flange and Down Into Rim Well.

5. Starting opposite the valve, use tire irons to lift top bead over rim flange and down into rim well. Care must be used to avoid pinching tube with tire irons. When bead is well started, lubricate remaining unmounted portion of tire bead and rim flange (Fig. 110).

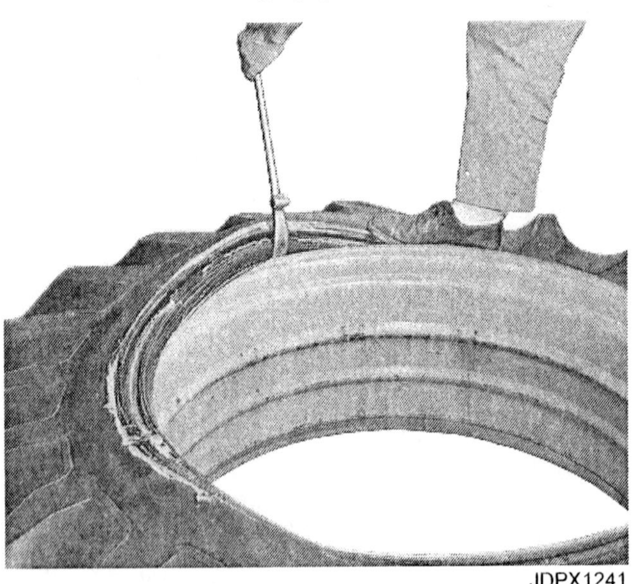

JDPX1241

Fig. 111 — Forcing Tire Bead Into Rim Well when Working Tire Bead Over Rim Flange

6. Tire bead must be forced into rim well when working tire bead over rim flange (Fig. 111) (lifting opposite side of bead may be required to do this). For small tire, stand on half of tire which is in rim well and work equally around both sides to spoon tire bead over rim flange until final section drops over at the tube valve. Do not attempt to pry large sections of bead over flange at one time.

JDPX1242

Fig. 112 — Two-Man Operation for Large Section Tires

With large section tires a second man may be required to hold tire bead in rim well, either with tire iron or by standing on the tire sidewall.

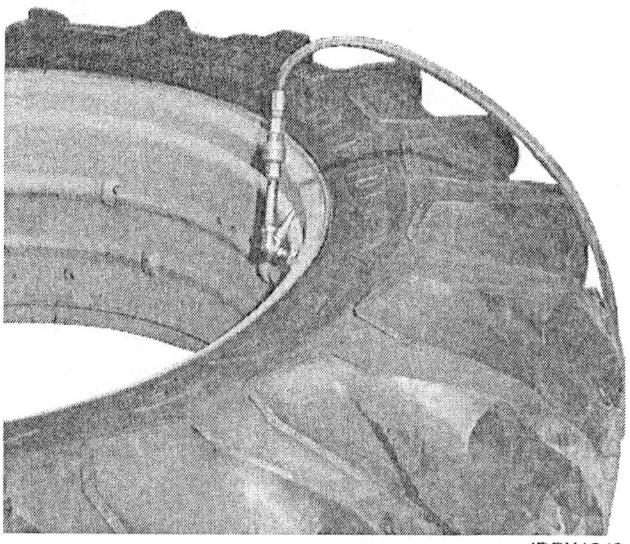

JDPX1243

Fig. 113 — Inflation of Centered Tire to Manufacturer's Specifications

CAUTION: Centering the tire is extremely important to prevent broken beads. Remote control inflation equipment should be used. Never stand over the assembly when inflating.

7. Center the tire on the rim and inflate to tire manufacturer's specifications.

8. Remove valve core housing and completely deflate.

9. Reinsert valve core housing and reinflate tire to recommended pressure.

NOTE: *If either bead fails to seat at recommended inflation, the tube may be pinched between tire bead and rim, or something else is interfering with proper mounting. Do not increase inflation pressure to seat beads, but remove valve core housing and completely deflate tube. Unseat both beads from the rim; relubricate both tire beads and rim bead seat areas. Reinstall valve core housing, and reinflate tube to tire manufacturer's specifications. Repeat process until both beads are properly seated.*

10. Reduce to operating pressure before putting tire in service.

Changing Large Tires for Off-the-Road, Industrial Equipment

Guidelines for mounting and demounting large tires for off-the-road industrial equipment tires are shown in this section. It is important that specific tire, wheel, and rim information is used to mount and demount these tires. See the manufacturers or manufacturer associations for this information. Proper tools, training, and procedures must be used. Mounting and demounting large off-the-road tires is a difficult job. Often, service must be performed in rough terrain far from a tire service center. Demounting and mounting a tire improperly is hazardous.

EQUIPMENT AND TOOLS

Tire changing steps will vary with different operators, but these tools are vital for changing large off-the-road tires:

- Heavy Equipment Jack
- Tire Tools and Irons
- Wheel Blocks
- Air Compressor
- Chains, Hoists, Supports

SAFETY TIPS FOR LARGE TIRES FOR OFF-THE-ROAD, INDUSTRIAL EQUIPMENT

- FOR YOUR SAFETY — Remember, an inflated tire and rim can be very dangerous; under pressure it packs the explosive force of TNT.

  A 10.00 - 20 12PR truck tire inflated to 75 psi has 46,510 Energy Foot-Pounds (63021 N•m) if it explodes, and could raise a 3000 pound (1362 kg) car to 15 feet (4.5 m).

  An 18-22.5 16PR tire with 75 psi (520 kPa) develops 103,789 Energy Foot-Pounds (140,634 N•m) and could hurl an 83-pound (37 kg) object over the Empire State Building.

  A 24.00 - 49 tire inflated to 75 psi (520 kPa) develops 354,260 Energy Foot-Pounds (480,022 N•m), and could lift a 134 lb. (60 kg) man one-half mile (0.8 km) into the air!

- Wear protective eyewear when using compressed air to prevent eye injuries caused by air-blown particles.
- Be sure to remove the valve core and exhaust **all** air from the tire. Check the valve stem by running a wire through the stem to be sure it is not plugged. A broken rim part under pressure could blow apart and kill you the moment you remove the lugs on a dual assembly. Remove valve cores from **both** tire assemblies. Exhaust ALL air from both tires before starting to loosen the lug nuts.
- Before jacking the equipment off the ground, take steps to keep it from rolling and falling. Regardless of how firm the ground appears, place hardwood blocks under the jack. Place blocks or stands under the equipment immediately after it has been jacked up.

Fig. 114 — Large Off-the-Road Tires Require Extra Safety Precautions

- Tire and wheel components are heavy. To avoid back and foot injuries, two people should handle large tires. If you have a hoist, lift truck or loader-equipped tractor, use it to handle tires.
- When using a cable or chain sling, stand clear. It might snap and lash out.
- Bead breakers and rams apply pressure to the bead flanges. **Keep your fingers clear.** Slant the bead breaker about 10 degrees to keep it firmly in place. If it slips off, it can be thrown with enough force to kill. **Always** stand to one side when you apply hydraulic pressure.
- Remove the bead seat band slowly to prevent it from dropping off and crushing your toes. Support the band on your thigh and roll it slowly to the ground. This will protect you back and toes.
- Never mix rim parts of different brands. Do not try to use the wrong size rim flange. Never use bent, chipped, or broken rim parts. The are dangerous and can cause injury.
- Never attempt to rework, weld, heat, or braze wheel parts. Always replace damaged parts with new parts of the same size, type, and make.

- Be very careful to clean all dirt and rust from the lock ring gutter. This is important to allow the lock ring to secure in its proper position. Inspect the rim base and lock ring gutter for cracks. Any cracked, damaged, or sprung rim bases or lock rings should be replaced.

- Be certain the lock ring is seated properly in the gutter completely around the rim before inflating the tire.

CAUTION: Do not attempt to seat any part by hammering or prying while the tire contains any inflation pressure. Explosive separation of the rim parts could occur and cause serious injury or death. If the parts are not seated properly, deflate the tire and correct the problem before proceeding.

- Never re-inflate or add inflation pressure to a tire that has been operated in a flat or near-flat condition (80 percent or less of recommended pressure). The tire may be damaged on the inside and may explode while you are adding air. The rim parts may also be damaged or dislodged and can explosively separate.

- Tire inflation equipment should have a filter to remove moisture. This prevents corrosion inside the tire.

- Place the tire and wheel assembly in an approved safety cage when inflating. If no cage is available, some other approved restraining device must be used. Use a pressure regulating valve, clip-on chuck and extension hose long enough to allow you to stand to one side and **NOT** in front of the tire assembly while inflating it.

- If the tire beads have not seated by the time the inflation pressure reaches the maximum pressure specified by the manufacturer, deflate the tire, reposition the tire on the rim, lubricate and reinflate the tire. **Do not exceed the maximum recommended inflation pressure in an attempt to seat the beads.**

- Air pressure is the only positive method to fully seat off-the-road tire beads. Operating the tire will not seat an improperly mounted tire, and it will result in damage to the beads and ultimate failure of the tire.

- Never, under any circumstances, introduce a flammable substance into a tire before, during or after mounting. Igniting this substance in an effort to seat the tire beads is extremely unsafe and could result in an explosion of the tire with a force sufficient to cause serious personal injury or death. This practice could also result in undetected damage to the tire or rim that could result in failure of the tire in operation.

- Spare tires mounted on demountable rims should only have enough air pressure to keep the rim parts in place. **Never transport a fully inflated tire.** Inflate the tire to the correct operating pressure at the jobsite using an approved safety cage or other suitable restraining device.

- Always contact the rim, wheel and tire manufacturers for specific procedures.

DEMOUNTING TUBE-TYPE TIRES FOR OFF-THE-ROAD, INDUSTRIAL EQUIPMENT

Contact tire, wheel, and rim manufacturers and associations for specific information.

IMPORTANT: **For safety suggestions when mounting and demounting tires, please refer to "Safety Tips For Large Tires For Off-The-Road, Industrial Equipment".**

1. Securely block all wheels other than the one which is being changed. Jack up the machine enough for the tire to clear the ground. For safety, block up under the axle so it won't fall if the jack slips.

2. Remove the valve core and exhaust ALL air pressure before starting to demount the tire. On dual tires, let the air out of both tires. Insert a wire into the valve to check for blockage. Always install the valve cap to protect the soft-metal threads of the valve during tire removal.

JDPX1245

Fig. 115 — Insert Special Tool in One of the Breaking Slots

JDPX1246

Fig. 116 — Twist the Special Tool

3. Place the flat hooded end of the special tool in one of the breaking slots between the seat band and the rim flange (Fig. 115). With a length of pipe slipped over the straight end of the tool for leverage, twist the tool in a circular direction as shown in Fig. 116.

4. Have a second person insert another tool between the bead seat and the side flange and, with the same twisting action, hold the amount of space gained with the first tool.

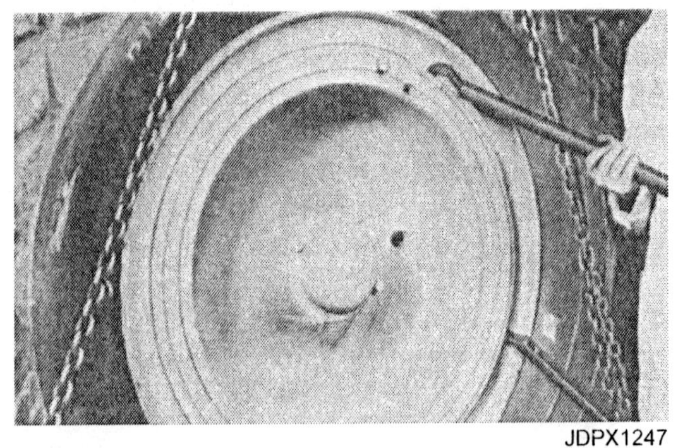
JDPX1247

Fig. 117 — Use Both Tools to Loosen Outer Bead

5. Move the first tool forward, around the rim, twist and follow up with the second tool (Fig. 117).

6. Continue in this way around the bead seat band until the outer bead is loose.

JDPX1248

Fig. 118 — Separate the Bead Seat Band from Lock Ring

7. Start separating the bead seat band from the lock ring by placing the tool in the gutter section between the ends of the lock ring and prying radially with the tool (Fig. 118).

8. Using two tools as in step 4, work completely around the tire.

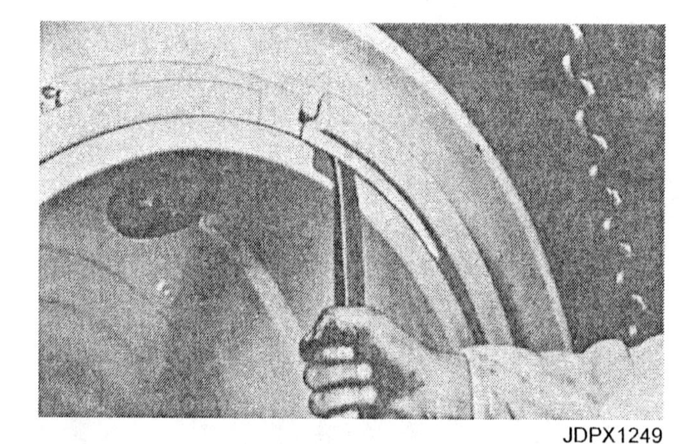

JDPX1249

Fig. 119 — Start Prying at the Prying Notch

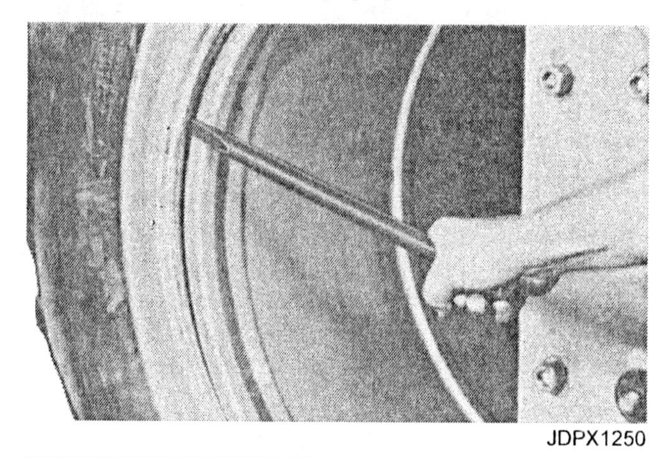

JDPX1250

Fig. 120 — Pry Out the Lock Ring with a Second Tire Iron

9. Pry out the lock ring by starting at the prying notch while holding with one tire iron (not shown) (Fig. 119) and working around the tire with a second tire iron (Fig. 120).

JDPX1251

Fig. 121 — Pry the Bead Seat Band Out Over the Gutter Section

10. After the lock ring has been removed, pry the bead seat band out over the gutter section (Fig. 121). Remove the bead seat band, together with the side flange, from the rim base. Use a hydraulic demounting tool to break the bead loose from the rim.

11. Similar prying notches are provided on the **inside** of the tapered bead rims, and the inside bead may be broken loose from the rim base in the same manner as the other bead.

    Another method is to place a small jack between some part of the vehicle and the tire. By extending the jack, pressure is exerted to force the tire bead off the rim taper. Work progressively around the tire to loosen the bead at all points.

12. Before working the tire off the rim, make certain the valve will clear the gutter section.

13. To work the tire off the rim, first force the tire outward as far as possible at the bottom, then lower the jack enough that the weight of the tire is resting on the ground. This will provide clearance at the top to permit you to force the tire out at the top by pushing or prying.

14. Hold the upper part of the tire as far out as possible, and raise the jack so the tire weight is resting on the top of the rim and clearance is gained at the top and the bottom. This allows the tire to be "walked" off the rim without having to work against the weight of the tire binding on the rim. A hoist or crane aids in this operation.

15. Completely remove the tire from the wheel and leave the tire in a vertical position, leaning against the vehicle or some other solid object.

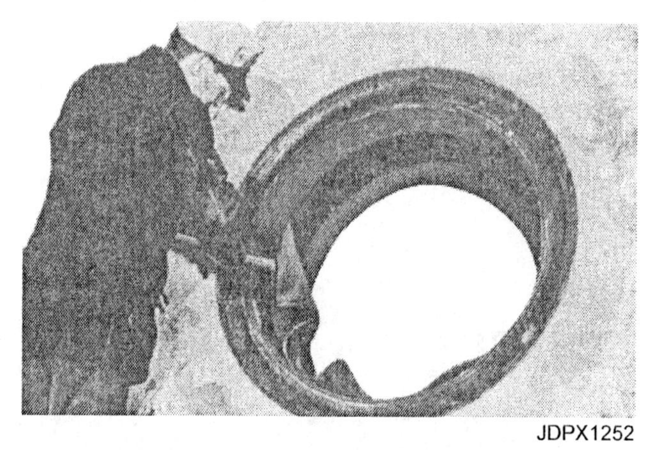

JDPX1252

Fig. 122 — Remove Flap from Tire

16. Remove the flap from the tire, using a tool with a rounded end to pry out and away from the beads (Fig. 122). On some types of tires, you may have to use a tire spreader or small jack to spread the beads and assist in removing the flap.

17. Remove the tube in the same manner, being careful not to pull the valve stem or enlarge any injuries.

MOUNTING TUBE-TYPE TIRES FOR OFF-THE-ROAD, INDUSTRIAL EQUIPMENT

Before remounting a tire on a wheel, clean the rust off the rim base and bead seats with a wire brush or by sand blasting. Check all rim parts for cracks. Replace unsafe parts with new ones. Paint these surfaces with one coat of zinc chromate, red lead, or a good grade of aluminum paint. Allow paint to dry thoroughly.

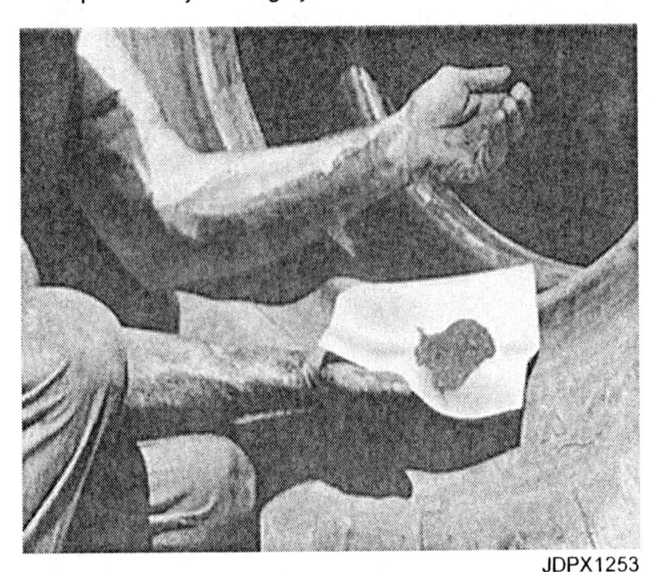

JDPX1253

Fig. 123 — Inspect and Clean Out the Inside of the Casing

1. Before installing the tube in the tire, inspect the casing carefully — inside and out — for breaks, bruises, nails, etc. Clean out all dirt and foreign matter from inside the casing (Fig. 123).

JDPX1254

Fig. 124 — Install the Tube in the Casing

2. Install the tube in the casing, starting at the bottom and working around the tire (Fig. 124).

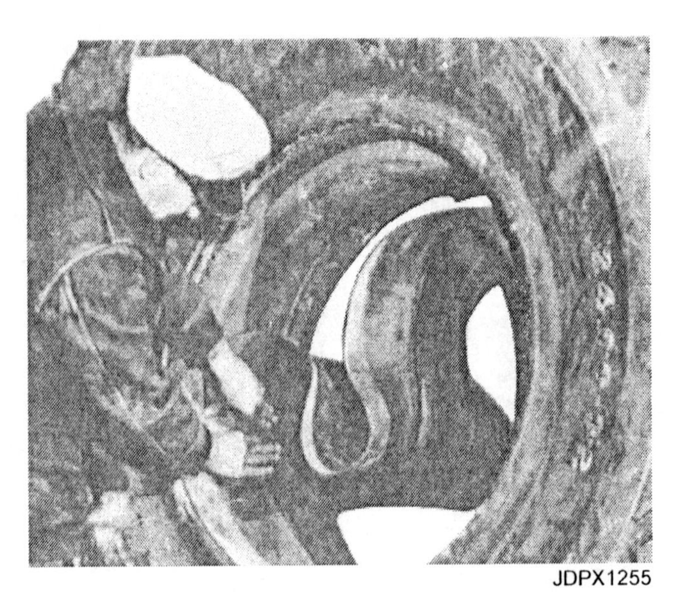

JDPX1255

Fig. 125 — Insert the Flap in the Casing

3. Insert the flap in the casing, making sure it is properly centered and smooth (Fig. 125). Rotate the tire as the flap is worked in, so the portion being inserted is at the bottom of the tire.

NOTE: In some cases, it may be necessary to spread the beads, using a spreader or small jack. Lubricate the beads with a vegetable oil soap in water solution or equivalent. DO NOT use petroleum-base or silicone lubricants.

JDPX1256

Fig. 126 — Align the Driving Lug with the Slot in the Rim (Rear Flange)

4. Install the rear flange on the rim base and align the driving lug with the slot in the rim (Fig. 126).

5. Hoist or roll the tire into position, making sure the valve stem aligns with the valve slot before sliding the tire onto

the wheel. If a hoist is not used, lower the wheel far enough to permit the top of the tire to be hooked over the top of the rim, then raise the wheel and "walk" the tire onto the wheel in the same manner as described for demounting.

JDPX1257

Fig. 127 — Hook Two of the Rack Tools into the Rim Gutter

6. Hook two of the rack tools into the rim gutter approximately 4 inches (10 cm) from either side of the valve slot (Fig. 127).

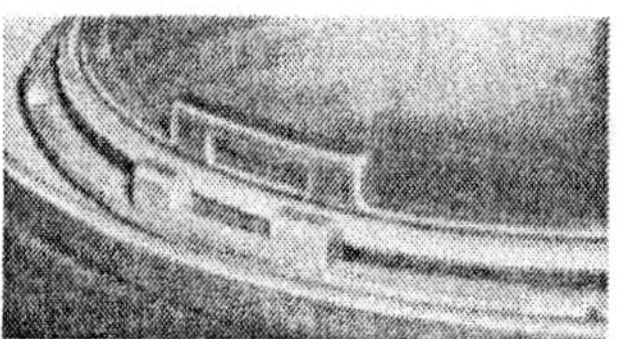

Align Driver Pockets in Bead Seat Band and Base.

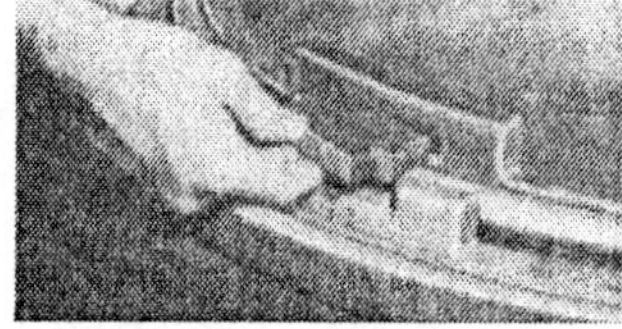

Insert Driving Slug into Driver Pocket on Base.

Make Sure All Parts Are Properly Aligned Before Inflation.

View Of Final Assembly.

JDPX1264

Fig. 128 — Typical Heavy-Duty Drive Application

7. Place the bead seat band in the outside flange, aligning the drive lug on the flange with the slot in the bead seat band (Fig. 128). Install the bead seat band and side flange in the rim base by hooking them over the two side tools and sliding on the rim. Be sure the driving lug on the bead seat band is aligned with the driver pocket in the rim base.

NOTE: The bead seat band will bind if cocked even slightly. If the band becomes wedged, do not hammer into place. Pull the band off the rim to loosen, then reposition it correctly on the rim base.

JDPX1258

Fig. 129 — Pry the Bead Seat Band into Position

8. Engage two of the bar tools in the proper notches of the rack tools and force the bead seat band into position by prying. While holding pressure against the band with one tool, slide the other rack and tool around the circumference of the rim, stopping approximately every foot (30 cm) to pry the bead seat band in place (Fig. 129). Follow up with the second rack and tool to hold pressure against the band and keep the clearance gained.

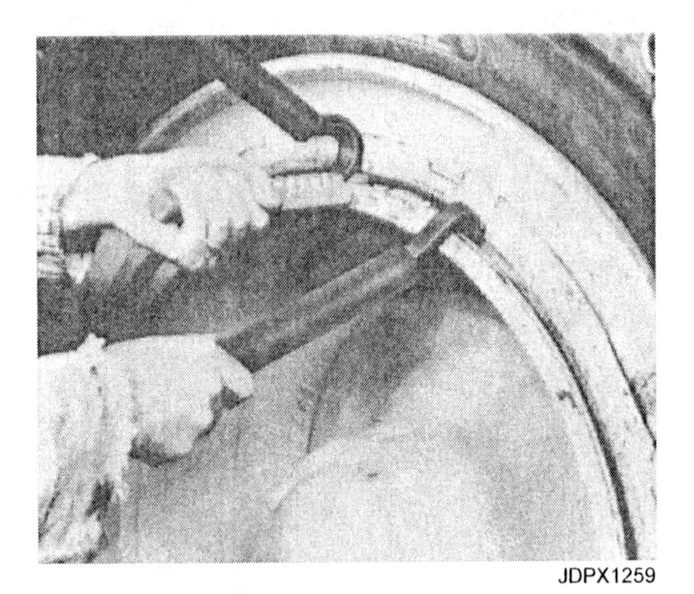

JDPX1259

Fig. 130 — Bar Tool Positioned in the Gutter for Additional Prying

9. After going around the rim once in this manner, it should be possible to hook one of the bar tools into the gutter for additional prying (Fig. 130). Continue to work around the rim, using the two tools without the racks until the bead seat has been forced 1 to 1-1/8 inches (25 to 28 mm) inside the gutter edge.

JDPX1261

Fig. 131 — Drive Tapered Wedge Between Bead Seat and Rim Base

10. Hold the band in this position and drive the tapered end of the bar tool, or any tapered wedge, between the bead seat and the rim base, directly under the break in the bead seat band (Fig. 131).

JDPX1262

Fig. 132 — Snap the Lock Ring into Place

11. Hook one end of the lock ring into the gutter. Pry or snap the lock ring into place using the tapered ends of the two bar tools (Fig. 132). When the lock ring is in place, remove the wedging tool from between the ends of the lock ring.

JDPX1263

Fig. 133 — Pry Bead Seat Band Out Over Edge of Lock Ring

12. Pry the bead seat band out over the edge of the lock ring, making certain that no "binding" occurs when this is done (Fig. 133).

CAUTION: Never start to inflate a tapered bead tire unless the bead seat band has been pried out over the lock ring. Place the assembly in a safety cage or other approved restraining device during inflation.

13. Check to make sure that the driving lugs on both the flanges are engaged.

14. Inflate the tire to the manufacturer's specifications to insure proper seating of the beads.

15. Completely deflate the tire. This will remove buckles and uneven stresses from the tube and flap

16. Reinflate tire to recommended pressure. This double inflation is necessary to prevent premature tube failures.

CAUTION: Use a clip-on chuck and extension hose long enough to allow you to stand well to one side and NOT in front of the assembly while inflating.

CAUTION: Failure to follow proper procedures when mounting a tire on a wheel or rim can produce an explosion which may result in serious injury or death. Do not attempt to mount a tire unless you have the proper equipment and expertise to perform the job.

When seating tire beads on rims or wheels, never exceed maximum inflation pressures specified by tire manufacturers for mounting tires. Inflation beyond this maximum pressure may break the bead, or even the rim, with dangerous explosive force.

17. Check newly mounted tires to be sure they are at proper inflation pressure before they go into service. Check again after the first two or three trips to eliminate the chance of a leaky valve or a bad tube ruining a good tire by underinflation.

DEMOUNTING TUBELESS TIRES FOR OFF-THE-ROAD, INDUSTRIAL EQUIPMENT

Contact tire, wheel, and rim manufacturers and associations for specific information.

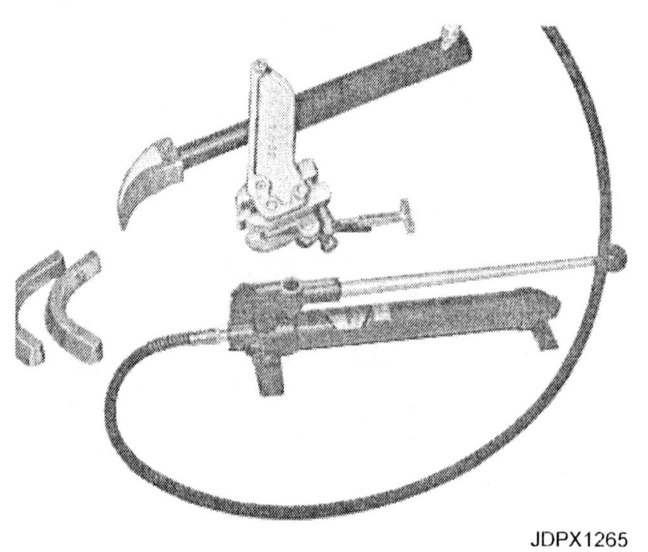

JDPX1265

Fig. 134 — Hydraulic Tire Remover

A hydraulic tire remover (Fig. 134) will normally be needed to demount large tubeless tires from their rims.

1. Prepare the unit by blocking the same way as for tube-type tires.

2. Remove the valve core and exhaust ALL air pressure before starting to demount the tire. Insert a wire into the valve to check for blockage. Always install the valve cap to protect the soft-metal threads of the valve during tire removal.

JDPX1266

Fig. 135 — Securely Tighten the Adjusting Screws

3. Securely tighten the adjusting screws at the bottom of the tire remover jaws (Fig. 135).

JDPX1267

Fig. 136 — Adjust the Jaw Assembly into a Right-angle Position to the Plane of the Flange

4. Set the hand screws against the lock ring and adjust until the jaw assembly is at a right-angle position to the plane of the flange (Fig. 136).

JDPX1268

Fig. 137 — Insert the Spade and Ram Assembly into Position

5. With the spade tip down, and the ram in the retracted position, insert the spade and ram assembly between the open sides of the frame. Place the spade top between the tire bead and the rim flange. Lift the ram until the trunnion engages the frame shoulder support and move the stop screw into the support ram (Fig. 137).

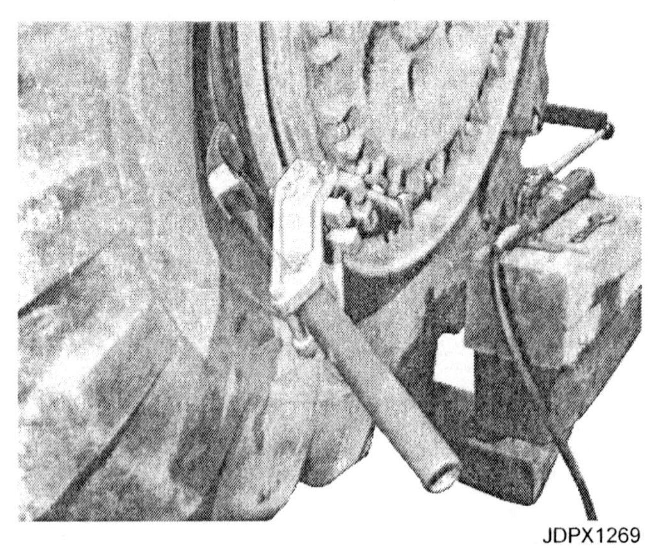

JDPX1269

Fig. 138 — Apply Pressure to the Ram and the Spade to Move the Tire Bead

6. Apply pressure to the ram and the spade with the hydraulic pump until the spade has moved the tire bead toward the center of the rim assembly far enough to permit placing a bead wedge between the bead and the flange on each side of the tool (Fig. 138). Release the pump pressure. Remove the spade ram assembly from the frame. Loosen the clamping jaw screws and remove the flange.

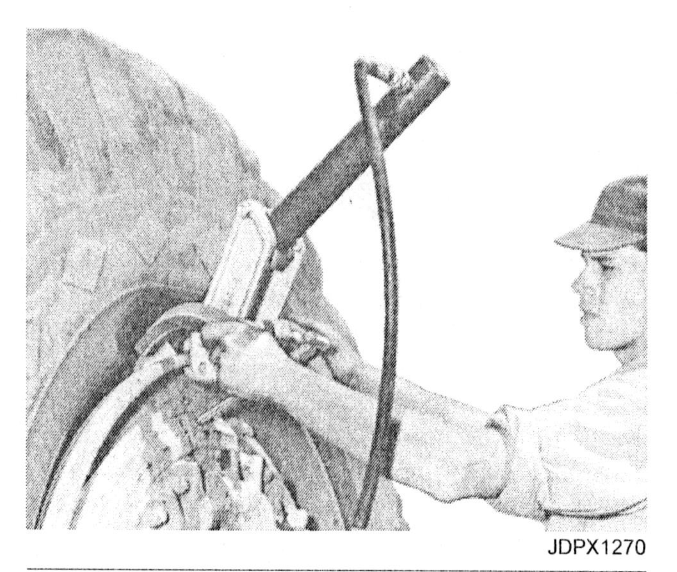

JDPX1270

Fig. 139 — Reposition the Spade and Ram Assembly and Repeat the Procedure

7. Move to a spot approximately 60 degrees from the first application (in either direction) and repeat the entire procedure (Fig. 139). Repeat until the tire is freed from the rim. Four or five applications may be required.

8. When using the hydraulic tire remover on a rim which has a 2 inch (5 cm) flange, you must inset the spade and ram assembly into the frame assembly and tighten before attaching it to the frame. This is necessary because of the small clearance between the flange and the tire.

JDPX1271

Fig. 140 — Hydraulic Bead Unseating Tool

CAUTION: To unseat the tire beads on large rims, use a hydraulic bead unseating tool (Fig. 140). This tool can only be used on rims which have pry bar slots or a continuous pry bar ledge. For detailed instructions, read the manufacturer's manual. Tools are available through your local rim or tire distributor.

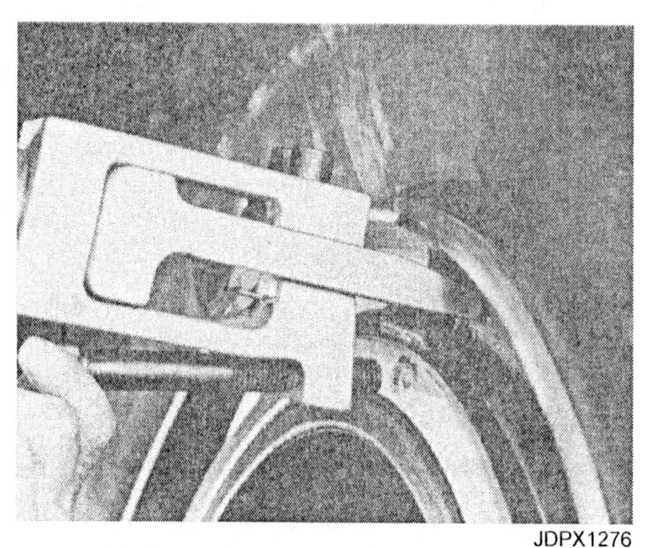

JDPX1276

Fig. 141 — Hydraulic Bead Unseating Tool

CAUTION: When using the hydraulic bead unseating tool, adjust the ram adjusting screw so the tool will remain perpendicular to the tire when under pressure (Fig. 141). Stand clear of the wheel assembly when applying hydraulic pressure to the bead unseating tool. If the tool is not properly seated, it can fly off with great force.

JDPX1272

Fig. 142 — Remove the Lock Ring Using a Pry Bar

9. Use a pry bar to remove the lock ring (Fig. 142), starting near the split end of the ring. Work the tool around the lock ring to remove the ring from the wheel.

JDPX1273

Fig. 143 — Pull the O-ring from the Rim Groove Using a Pry Bar

10. Depress the tire and the bead seat band using a pry bar or a hydraulic demounting tool to gain access to the O-ring. Insert a pry bar under the O-ring and pull it from the rim groove (Fig. 143). A new O-ring should always be used when remounting the tire.

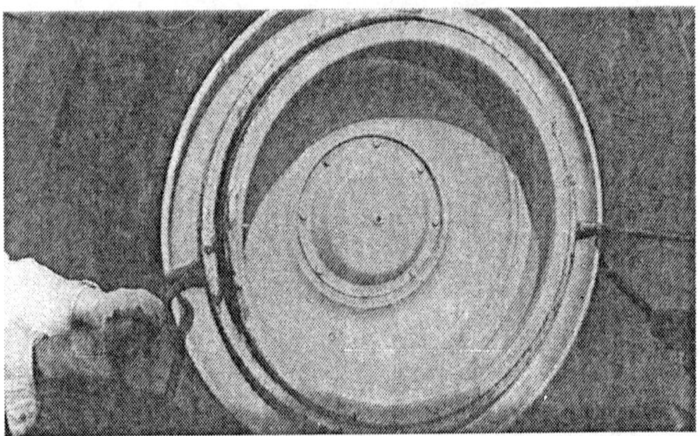

JDPX1274

Fig. 144 — Pry the Bead Seat Band Loose Using a Pry Bar

11. Insert a pry bar under the rim flange and pry the bead seat band loose (Fig. 144). Use pry bars or a hoist to remove bead seat band from the rim. Remove the front flange from the rim.

JDPX1275

Fig. 145 — Unseat the Back Tire Bead Using a Short Hydraulic Cylinder

12. If the wheel assembly is off the machine, turn the assembly over and use the same procedure to unseat the back tire bead. If the wheel assembly is mounted on the machine, use a hydraulic bead unseating tool or a short hydraulic cylinder positioned between the machine frame and the back flange to unseat the back tire bead (Fig. 145).

13. Use a hoist and sling or a suitable tire handler to remove the tire from the wheel.

CAUTION: Stand clear when using a cable or chain sling. It can snap and lash out.

MOUNTING TUBELESS TIRES FOR OFF-THE-ROAD, INDUSTRIAL EQUIPMENT

It is important that all rim parts be examined for damage and corrosion. Rims should be clean, especially the groove in which the O-ring gasket seats. A new O-ring should always be used. Lubricate the O-ring with vegetable oil soap in water solution or other approved rubber lubricant.

Tubeless tire changing tools can be ordered through several sources. Tools required are off-the-road tire irons and hydraulic bead breaking tools. For specific rim assembly, contact the rim manufacturer.

Guidelines

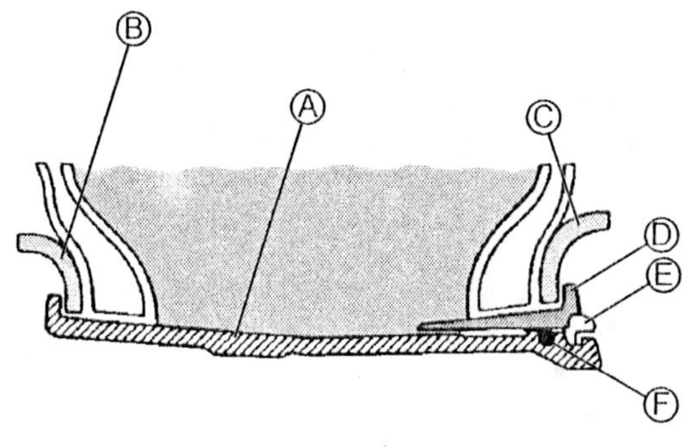

Fig. 146 — Tubeless Tire for Off-the-Road, Industrial Equipment (Mounted on a Rim and Inflated)

The cross section of a large tubeless tire mounted on a rim and inflated is shown in Fig. 146. Mounting tubeless tires is the same as mounting tube-type tires except for the following:

- Rims for tube-type and tubeless tires have similar parts, except a sealing rubber O-ring (F) which goes between the tapered bead seat band (D) and the rim base (A).
- The rim base (A) does not have a rim slot in it.
- The tapered bead seat band (D) is one-piece and is not split as is the tapered band for tube-type rims.

Before mounting or assembling the rim, all parts should be wire brushed thoroughly. This includes the entire rim base, the back flange seat, the O-ring groove and the lock ring gutter. Paint or coat all parts with a rust inhibitor.

Lubricate the O-ring groove, and the area behind it for approximately 2 inches (5 cm), with vegetable oil soap in a water solution or other approved rubber lubricant. Do not use petroleum or silicone base lubricants. Make sure the valve stem is properly installed in the rim base.

Tubeless tires can be mounted in two ways:

- Vertical mounting (wheel on the vehicle)
- Horizontal mounting (wheel off the vehicle)

Vertical Mounting of Tubeless Tires for Off-the-Road, Industrial Equipment — Guidelines

1. Clean the rear flange and install it on the rim base. Push the flange to the back shoulder of the rim base. Make certain the drive lug on the flange properly engages the slot in the rim base.

Fig. 147 — Use a Hoist to Support and Install Flange on Large Wheels

2. Use a hoist and sling to position the tire on the rim base. Force the tire back as far as possible against the back flange. Use a hoist to install the outside rim flange on the rim base (Fig. 147).

JDPX1279

Fig. 148 — Use a Hoist to Install Bead Seat Band on Large Wheels

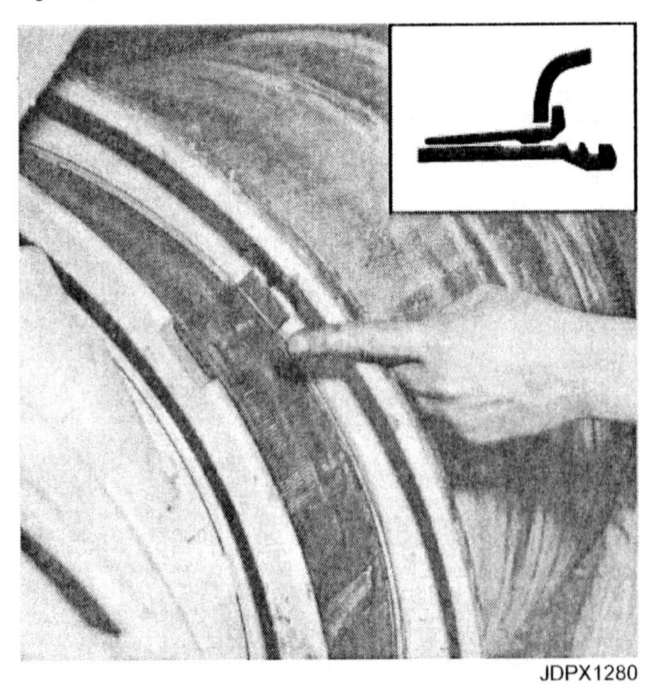

JDPX1280

Fig. 149 — Installing Bead Seat Band on Rim Base

3. Install the tapered bead seat band over the rim base (Fig. 148), forcing it between the flange and rim base. Be sure to align the drive lug slot in the tapered bead seat band with the drive lug on the rim base and flange (if present) as shown in Fig. 149.

NOTE: *The bead seat band will bind if cocked even slightly. If the band becomes wedged, DO NOT try to hammer it in place. Remove the band from the rim and reposition it correctly on the rim base.*

4. Use tire irons to force the outside flange assembly, the tapered bead seat band and the outside tire bead in far enough to clear the ring gutters.

JDPX1281

Fig. 150 — Make Sure Drive Lug on Lock Ring Fits in Slot in Rim Base

5. With the outer flange and tapered bead seat band held in the above position, insert the lock ring in the outside gutter. Top the lock ring into place with a soft hammer, making sure the drive lug on the lock ring is aligned with the slot on the rim base (Fig. 150).

For specific rim assembly, contact the rim manufacturer.

6. Be sure that all slots and drive lugs are properly aligned. If it is necessary to move any of the parts, use only a lead or brass hammer to tap the parts.

7. Lubricate the O-ring gasket with vegetable oil soap in a water solution or other approved rubber lubricant. *Always use a new O-ring when mounting a tire. Never re-use an old O-ring, as it becomes seriously distorted in*

making a seal. An old seal ring could prevent a perfect seal.

8. Lubricate the O-ring groove area with rubber lubricant. Place the bottom of the O-ring over the lock ring into the rim gutter without twisting or rolling the O-ring. Spread the O-ring over the rim base and lock ring, and snap into place in the rim groove (Fig. 150, upper right inset). *Be careful to place the O-ring in the rim groove without rolling or twisting. Lubricate the O-ring area again by applying rubber lubricant with a brush.*

Fig. 151 — Positioning O-Ring, Bead Seat Band and Outside Flange

9. Pull the tapered bead seat band and the outside flange out over the O-ring as tightly as possible against the lock ring (Fig. 151). At this point, make sure all the rim parts are correctly assembled before inflating the tire.

10. In order to seat the tire beads, position the rim and tire in a safety cage or use some other approved restraining device during tire inflation. Inflate the tire to the pressure recommended by the tire manufacturer, using a clip-on chuck with an in-line pressure gauge and control valve. *Always stand away from the tire assembly during inflation.*

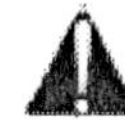

CAUTION: Failure to follow the proper procedures when mounting a tire on a wheel or rim can produce an explosion which may result in serious injury or death. Do not attempt to mount a tire unless you have the proper equipment and expertise to perform the job.

When seating the tire beads on rims or wheels, never exceed the maximum inflation pressures specified by tire manufacturers for mounting tires. Inflation beyond this maximum pressure may break the bead, or even the rim, with dangerous explosive force.

11. Check all the rim components for proper seating. If necessary, deflate the tire and adjust components.

CAUTION: Do not attempt to seat any part by hammering or prying while the tire contains any inflation pressure. This could cause an explosive separation of the tire components, resulting in possible serious injury or death.

12. Adjust to proper operating pressure. Check the tire assembly for air pressure loss. Do not put a tire into service that has a slow leak.

Horizontal Mounting of Tubeless Tires for Off-the-Road, Industrial Equipment — Guidelines

Horizontal mounting is the same as vertical mounting except for two additional steps:

For specific rim assembly, contact the rim manufacturer.

JDPX1283

Fig. 152 — Place Rim Base on Blocks, Gutter Side Up, for Horizontal Mounting

1. Lay the rim base on wood or metal blocks, gutter side up (Fig. 152). The rim should be off the floor enough for the tire to rest on the rim without touching the floor. The blocks should not extend more than 1/2 inch (13 mm) beyond the rim base to prevent interference with the back rim flange.

2. Some tires may require slight lifting to contact the upper bead seat band. Use a hoist and chain sling or a tire handler to lift the tire. Position an approved restraining device around the tire, then inflate using an extension hose with a clip-on chuck and air pressure regulator. Always stand to one side and NOT in front of the assembly while inflating.

Summary: Tires

We have tried to describe all the common types of tires used on today's farm and industrial machines.

After this review, it should be obvious that, unless you have a very large operation, tire service and repairs are better left to those in the tire business.

Without proper tire equipment, which is expensive, it is usually unsafe and impractical to make satisfactory repairs.

Nevertheless, you should know how tires are constructed and rated, how they are inflated and ballasted, how they fail, and how to perform basic tire maintenance.

Test Yourself

QUESTIONS

1. Which of the items below is not true of off-the-road tires (when compared to over-the-road tires)?

   a. Operate at lower speed

   b. Generate less heat

   c. Receive less shocks

2. If a tire size is "24.5 - 32," what do the two numbers tell you?

3. Which way should the V-treads face on a farm tractor drive tire when viewed from the front?

   a. Upward (Λ)

   b. Downward (V)

4. True or false? "Tires are designed to operate with some sidewall deflection or bulge."

5. Will underinflated tires wear more on the sides or the center of the tread?

6. Match each item below with the correct answer at the right:

a. Too much ballast	1. Tire tread pattern wiped out
b. Too little ballast	2. Tire tread pattern visible but broken or shifted
c. Proper ballast	3. Tire tread pattern a distinct mold of tire

7. What is the proper tire slippage for a farm tractor under load in the field?

   a. No slip

   b. 10-15 percent slip

   c. 20-25 percent slip

8. When filling a tire with liquid ballast, how full should it be?

   a. 75%

   b. 90%

   c. 100%

9. If rubber tires are operated for a long period on hard surfaces, should inflation pressure be at the highest or lowest level recommended?

10. How should minor cuts or snags on the rubber of the tire lugs be treated?

11. True or false? "Hot patches make a better repair for punctured tires than cold patches."

12. How should tires be stacked up for storage?

13. After mounting a tube-type tire, why should it be inflated and then deflated again?

14. True or false? "The safest place to stand while inflating a tire is over it."

ANSWERS

1. c — Receive less shocks.

2. "24.5" tells the width of the tire in inches, while "32" gives the diameter of the rim.

3. b — Downward.

4. True.

5. Sides of the tread will wear more.

6. a — 3; b — 1; c — 2.

7. b — 10-15 percent slip.

8. a — 75%.

9. Highest.

10. Bevel out the cut or hole with a knife (to prevent more tearing or picking up of stones).

11. True.

12. They should **not** be stacked as this will distort them. The best way is to mount and store them in a vertical position.

13. To prevent buckling or pinching of the tube.

14. False — Stand to the side, away from the tire.

METAL TRACKS

2

The track and undercarriage may represent 20 percent of the purchase cost and 50 percent of the service cost of a crawler tractor. This makes good operation and maintenance of tracks a major concern of owners and service shops.

Tracks are built rugged for use in muddy, rocky, or loose soil. However, reckless operation can put violent shocks on track parts and break them. Not keeping tracks adjusted and in good repair can do equal harm.

Let's look at the track and undercarriage, how it works and how it can fail:

Operation

JDPX1335

Fig. 1 — Tracks Are Like Roller Chains with ShoesTracks are like roller chains with shoes (Fig. 1).

Each track system has seven basic parts:

- Track — endless chain
- Drive Sprocket — support and align track
- Rollers — support and align track
- Front Idler — maintains track tension
- Tension Mechanism — adjusts track tension
- Roller Guards — protect rollers and guide track
- Frame — supports all track parts

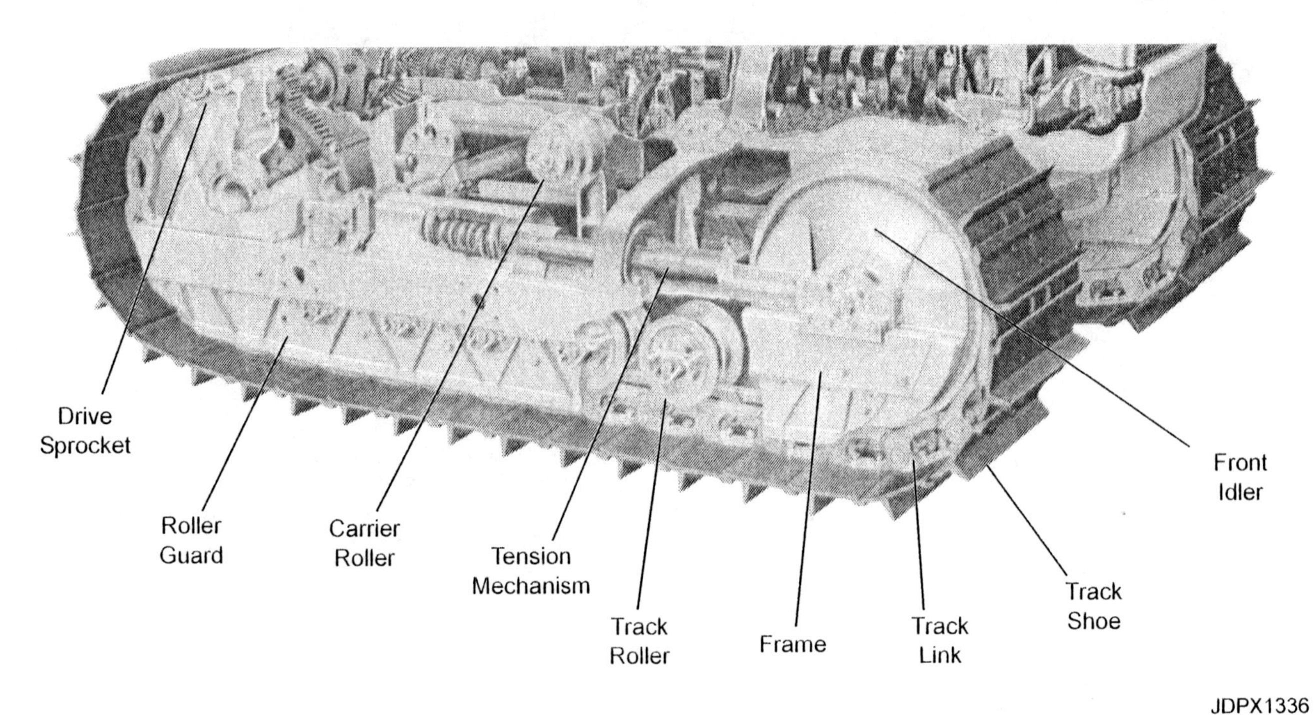

Fig. 2 — Complete Track or Undercarriage of Crawler Tractor

All these parts together make up the *undercarriage* of a crawler tractor (Fig. 2). One set on each side of the machine, driven by the tractor's final drives, supports and moves the complete machine.

Let's look at the function of each part.

TRACK

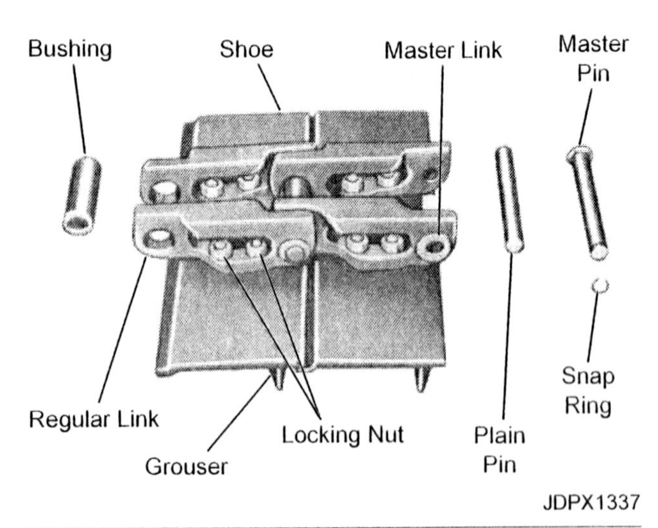

Fig. 3 — Track Links and Shoes

The track is an endless chain made up of links, bushings, pins, and shoes as shown in Fig. 3.

Each section of the track is a pair of links fastened together with a bushing at one end and a pin at the other. Both the pin and bushing are "press fit" into the links so the section won't work apart. The pin fits inside a bushing to hold the next pair of links.

Each pin and bushing set acts as a hinge as the chain links travel around the drive sprocket and front idler. To resist the wear caused by the resulting friction as the pin rotates in the bushing, the pins and bushings are heat-treated and induction hardened, and the bushings are closely mated to the pins to provide a true bearing surface. Sealed and lubricated track chain also has a lubricant sealed in between the bushings and pins, which virtually eliminates internal wear of the bushings.

TRACK (CONTINUED)

1. Rotate track until split link is positioned in front of idler.
2. To prevent track from falling, block track as shown.
3. Remove four cap screws to remove track shoe and disassemble split link.
4. Attach chain to track and support track chain with hoist to avoid damage to split link.
5. Start machine and slowly rotate drive sprocket in reverse direction to move track off drive sprocket.

IMPORTANT: Track chain must be installed under track frame with wide end of links toward rear of machine or accelerated wear to track chain will occur.

6. Install track chain by putting chain under track frame with the wide end of links toward rear of machine.
7. Block track shoes so master split link is positioned directly over idler.

JDPX8180

Fig. 4 — Track Shoe

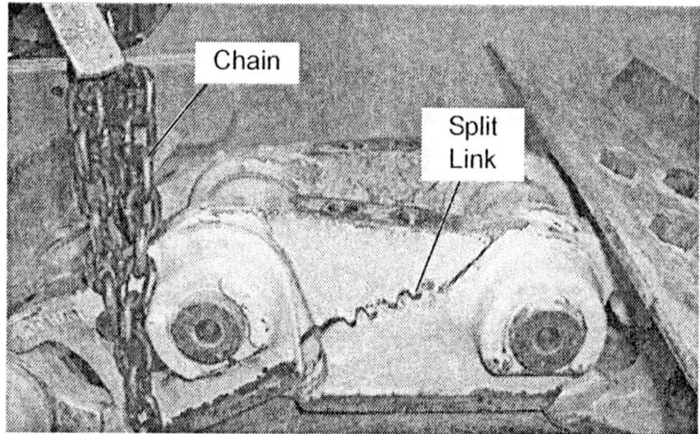

JDPX8181

Fig. 5 — Split Chain Link

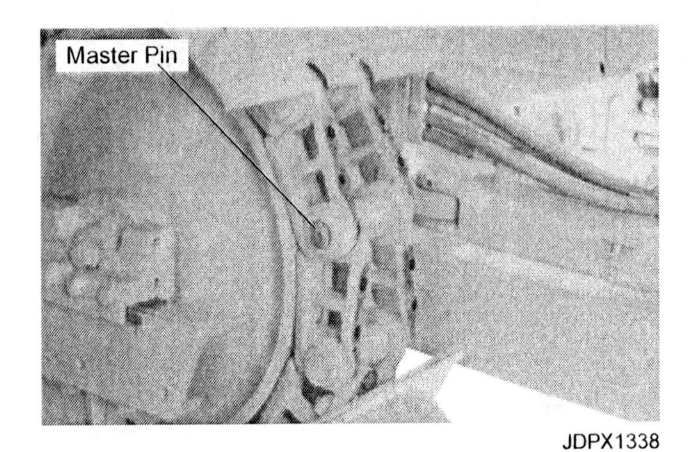

Fig. 6 — Track Master Pin Is Identified by a Drill Point in the End of the Pin

Each track chain has one section known as a "master link" that can be disconnected to provide a method of separating the track. The conventional master link consists of a special master pin and bushing which differ from the other pins and bushings. The master bushing is shorter than normal and the master pin is smaller in diameter for easier removal and installation. The master pin is identified by a drill point in the end of the pin (Fig. 6).

Fig. 7 — Split Master Link Is a Two-Piece Link

The split master link (Fig. 7) is a two-piece link assembly that is held in the locked position by the track shoe and special bolts. To disconnect the split link, it is only necessary to remove the track shoe bolts. The split link eliminates the need for special equipment to split the track chain assembly.

The split master link uses the same pins and bushings as other links of the track chain. This eliminates the special pins and bushings used on conventional master links, providing uniform wear life of all the chain joints.

Several different design split master links are produced by the manufacturers of undercarriage components. The split master link can be used with either conventional or sealed and lubricated pins and bushings.

Track shoes are formed metal plates that are bolted to each link of the track chain. The track shoes make contact with the ground to provide traction. The "grip" or grouser on the shoes will vary for use on different terrains (Fig. 8).

Types of Track Shoes

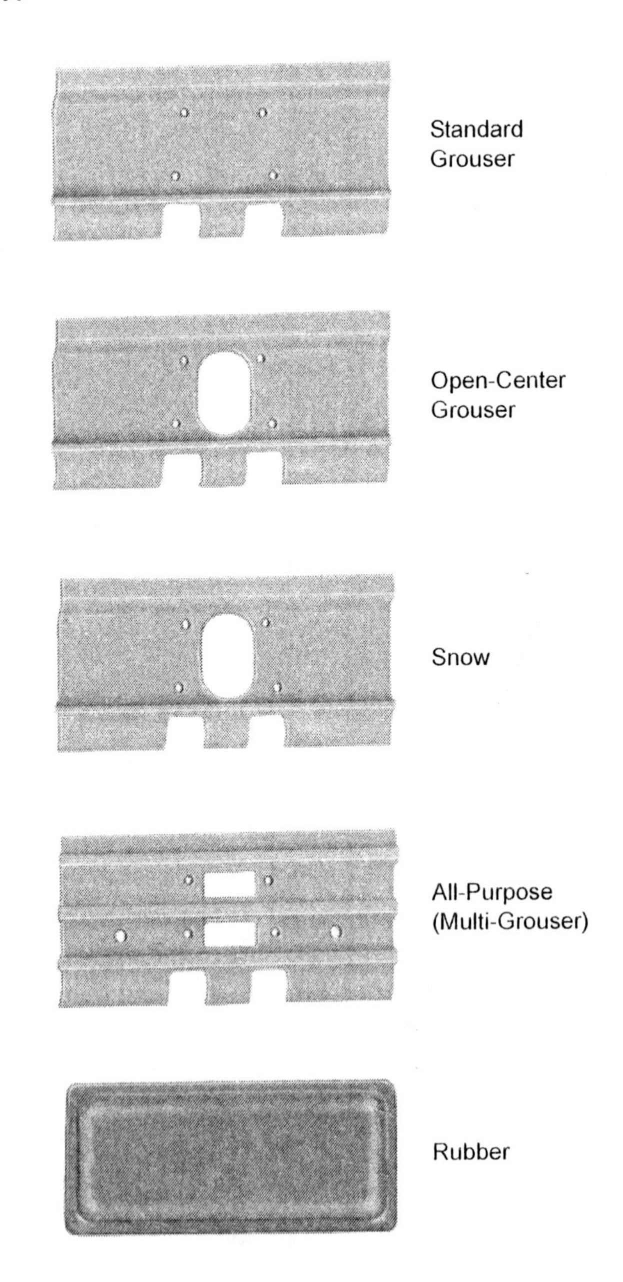

Fig. 8 — Common Track Shoes

Standard grouser shoes are widely used on dozer and drawbar machines and provide excellent service for most conditions. They are designed to work in clay, silt, loam, and gravel.

Open-center grouser shoes are a variation which provide openings for mud to squeeze out and prevent build-ups.

Snow shoes have similar slots which prevent snow and ice from packing in the tracks.

All-purpose (multi-grouser) shoes have small, low grousers which make the track more maneuverable. They also have self-cleaning openings. These shoes perform well on hard, smooth surfaces as well as in muddy or soft conditions. They are commonly used on loader units.

Rubber shoes are excellent for use on floors or other smooth surfaces that might be damaged by rough tracks. The wearing life is fairly short compared to steel shoes.

Special track shoes are also available for extreme or unusual conditions. For example, some shoes are extremely hard for use in abrasive conditions. Another shoe for rocky terrain will "work harden" as it is used.

Wider track shoes are sometimes available for special conditions such as loose snow. But remember, for maximum service life, track shoes should never be wider than necessary for the desired flotation.

Types of Track Chains

Track chains are made in two designs:

- Flush-type
- Interlocking type

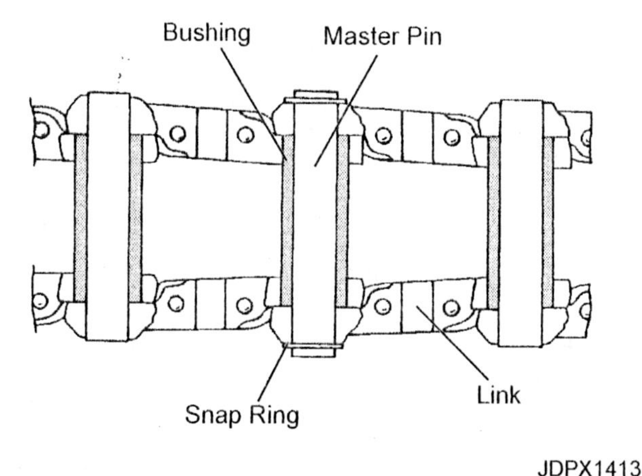

Fig. 9 — Flush-Type Track Chain

Flush-type chains have their bushings flush with the ends of the links (Fig. 9).

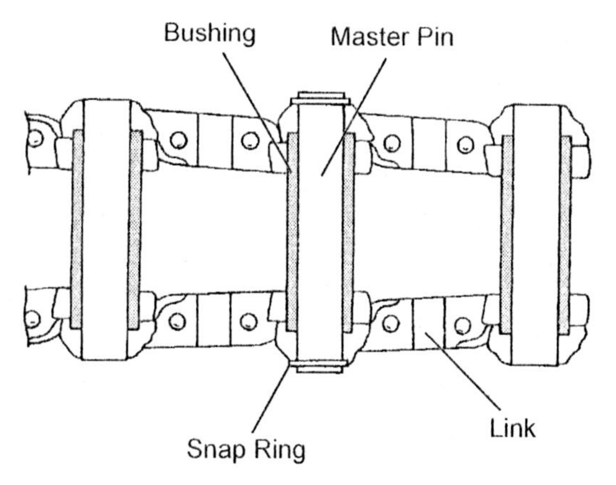

Fig. 10 — Interlocking-Type Track Chain

Interlocking-type chains have their bushings extending into a counterbore in each of the links (Fig. 10). This provides more protection against wear than does the flush-type chain because less abrasive material gets into the area between the pin and bushing. They also have more area of contact between the pins and bushings, which increases the wear life. Interlocking-type chains may also be equipped with seals, located in the link counterbores, to further reduce the amount of abrasive material that can enter the area between the pin and bushing.

Track Seals (Conventional Track)

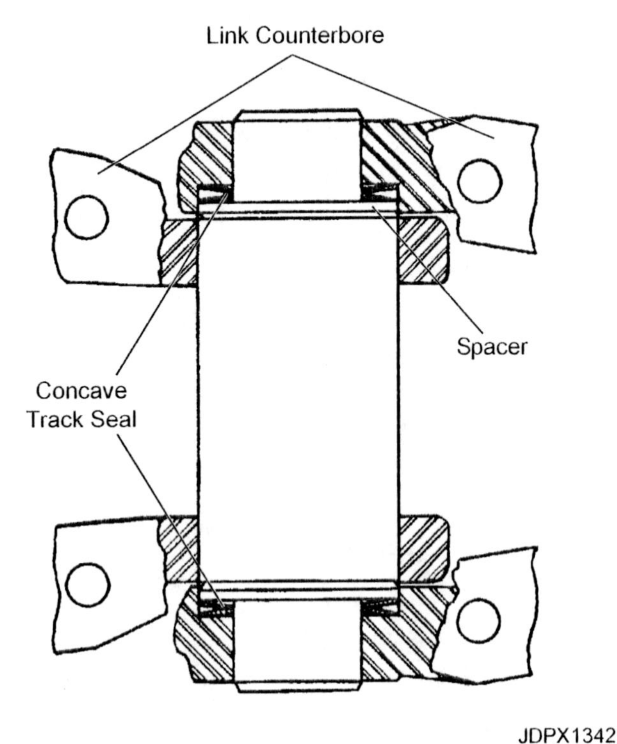

Fig. 11 — Two Concave Seals Are Located in the Track Link Counterbore on a Sealed (Non-Lubricated) Track

Two concave track seals, positioned with their inside diameters touching, are located between the track link counterbores and the end of the bushings on each side of the track link assembly (Fig. 11). The inner seal contacts either the end of the bushing or a spacer, which is located between the seal and the end of the bushing. The outer seal contacts the link counterbore.

The track seals reduce the amount of dirt and other abrasive material that can enter the space between the pin and bushing. The seals extend the life of the pins and bushings and reduce wear in the link counterbore area.

Track Seals (Lubricated Track)

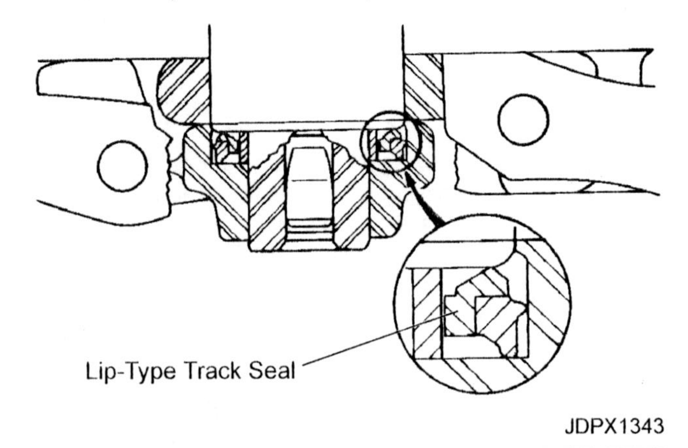

Fig. 12 — Lip-Type Seals Are Used on a Sealed and Lubricated Track

Lip-type track seals are used in a sealed and lubricated track chain (Fig. 12). The seals are located in the track link counterbores and form a positive seal against the counterbore, the end of the bushings and the outer diameter of the pins.

The lip-type seals provide a more effective seal than the concave seals that are used in a conventional track chain. The lip-type seals not only prevent the entry of abrasive material into the space between the track bushings and pins, as in the sealed conventional chain, but they also seal in lubricant between the bushings and pins.

A sealed and lubricated track initially costs more than a sealed conventional track. However, over the life of the undercarriage, the operating cost for the sealed and lubricated track should be lower than that of the conventional track. This is due to the virtual elimination of the internal wear of the pins and bushings that occurs on a conventional track.

As a result, all related wear of the undercarriage is greatly reduced, including sprocket wear and roller flange wear. The external bushing wear life is also increased.

A sealed and lubricated track also eliminates much of the loud, squealing noise created by the contact of the metal-to-metal parts of the conventional track chain. Another benefit of the reduced friction provided by the lubricated track chain is that the machine is capable of delivering more horsepower to the ground to perform more work when compared with a comparable machine equipped with a conventional non-lubricated track.

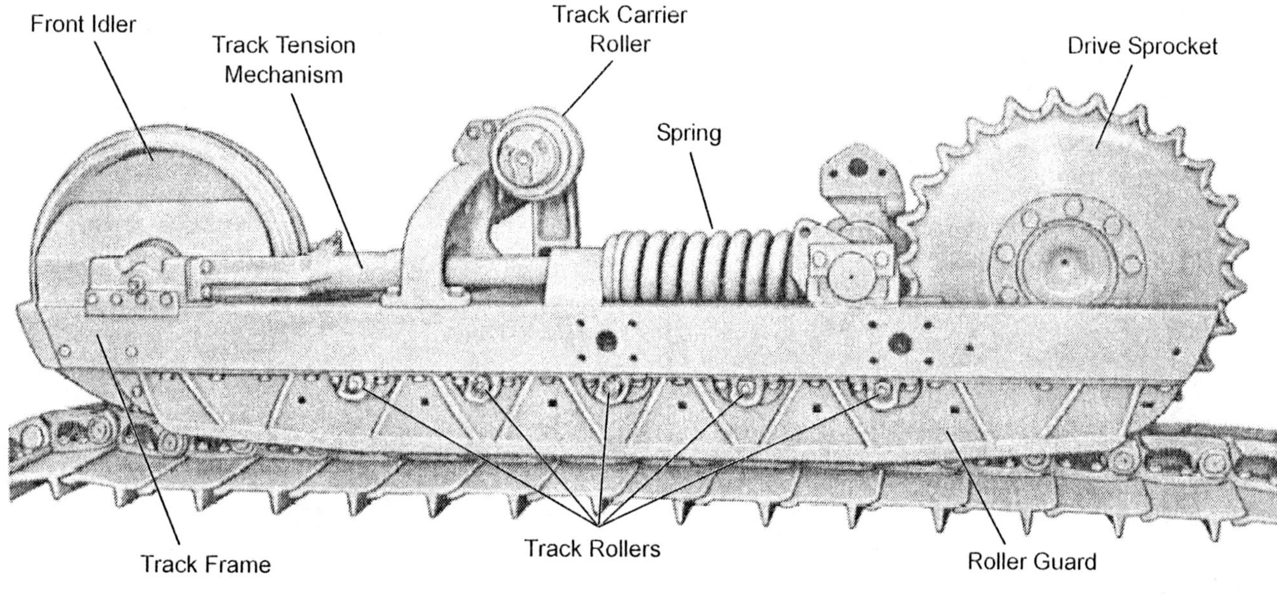

Fig. 13 — Track and Undercarriage of Crawler Tractor

DRIVE SPROCKET

The drive sprocket is driven by the tractor final drive and transmits power to the track (Fig. 13). The "hunting tooth" design of the sprocket produces even wear on the sprocket teeth. Every other tooth engages a track bushing and carries it around the sprocket. Since the sprocket has an odd number of teeth, each tooth catches a bushing every other time around.

ROLLERS

In Fig. 13, two types of rollers are shown. The carrier roller supports the weight of the upper portion of the track. It prevents the track from sagging, which causes excessive track whipping.

The track rollers mounted on the bottom of the frame support much of the machine's weight. They are small and close together to provide even pressure on the track as it revolves and contacts the ground. The weight is then distributed evenly over the entire bottom of the track. This gives the track its great traction and flotation.

FRONT IDLER

The front idler supports the track as it revolves (Fig. 13). A tension spring provides constant but flexible tension on the track thought the tension mechanism.

TENSION MECHANISM

The tension mechanism (Fig. 13) is usually a recoil spring and adjustable rod which can be turned manually or actuated hydraulically to tighten or loosen the tracks. The tension mechanism also absorbs impacts to the rail by use of a spring.

ROLLER GUARDS

The roller guards (Fig. 13) protect the rollers from rocks and other obstructions. They also help keep the track free of dirt and debris by preventing it from entering the track chain.

Another vital job of the guards is to guide the tracks and prevent twisting and buckling.

FRAME

In Fig. 13, notice that the frame supports and holds the track components in place. The front idler slides on the frame, the rollers are mounted on it, and the roller guards are attached to it. However, the drive sprocket is not supported by the frame. It is supported by its own axle from the final drive and steering clutch housing. On some machines, however, the outer end of the drive sprocket axle is supported by the frame.

TIPS FOR CRAWLER OPERATORS

These tips for the machine operator can extend track life:

- Avoid reckless operation over rough, rocky ground.
- Don't spin the tracks — apply loads gradually.
- Slow down — especially over rough terrain.
- Avoid high-speed operation in reverse.
- Stay off concrete and other hard surfaces when possible.
- Keep roller guards in place — especially when working in rock.
- Clean out packed mud after operating in muddy conditions.
- Check track tension more often when operating in mud, snow, or sand.
- Don't park on the side of a hill as this puts a strain on roller seals.
- In freezing weather, park on a hard surface to avoid freezing tracks to ground.
- Check tightness of all track bolts periodically.
- Inspect the track regularly.

Cleaning, Inspecting, and Lubricating Tracks

Fig. 14 — Mud Caked on Tracks Means Trouble

Muddy tracks mean trouble if not cleaned before the mud hardens. They cannot operate properly when caked with rock-hard mud (Fig. 14).

Rocks packed into the track can also cause a tight track, resulting in extra wear and stress on the track. The roller guards help keep out rocks, but the tracks should be checked for rocks periodically.

A good daily inspection of the track takes only a few minutes but can prevent major problems before they happen.

Inspect the complete track and undercarriage for misaligned, loose, or missing parts. Use a torque wrench to tighten loose bolts, especially on track shoes.

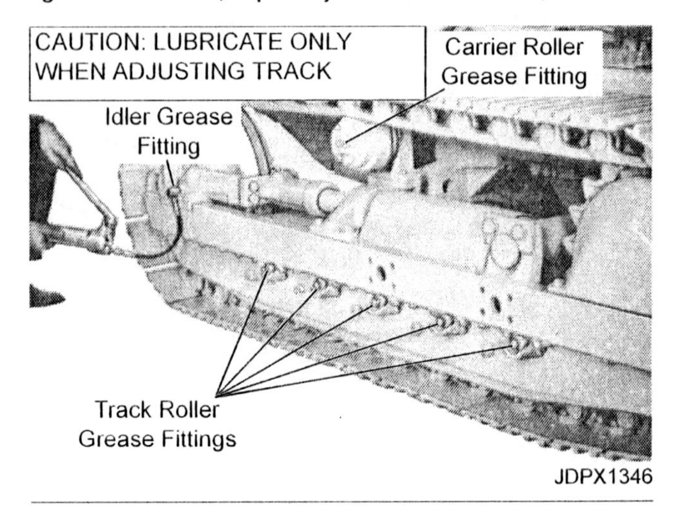

Fig. 15 — Lubricate the Track (If Recommended)

The hardened track pins and bushings are lubricated internally on a sealed and lubricated track, and require no periodic maintenance. However, the track rollers, carrier roller, and idler may have grease fittings (Fig. 15). At the correct intervals, clean the fittings and apply the recommended grease. A special low-pressure grease gun may be supplied with the machine. Normally, grease should be applied only until resistance is felt on the lever of the special grease gun.

IMPORTANT: **If the track has a hydraulic tension adjuster, never apply grease to this fitting except when adjusting the track tension. Doing so will overtighten the track.**

Some newer tracks have "lifetime" seals which are filled with grease during assembly and need not be lubricated.

MEASURING EXCAVATOR, FELLER-BUNCHER, AND LOGGERS TRACK SAG

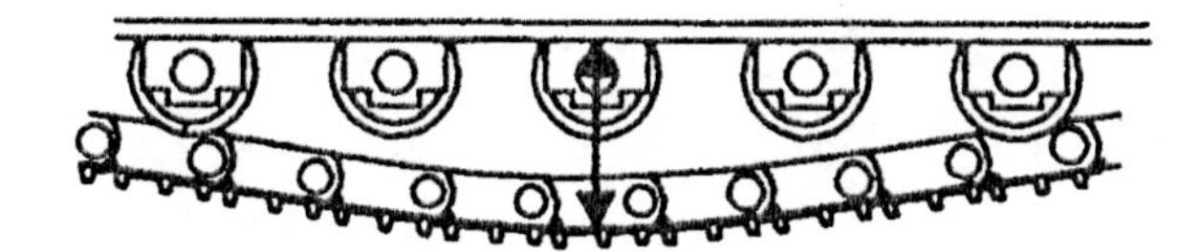

Fig. 16 — Measuring Track Sag

Check track sag by raising the undercarriage off the ground at center of track.

Excavators — Measure from the center of track frame between the track shoe and track roller mounting surface.

Feller-Bunchers and Loggers— Measure from the trackshoe to the outside most lower edge of track guard.

Decreasing Track Sag

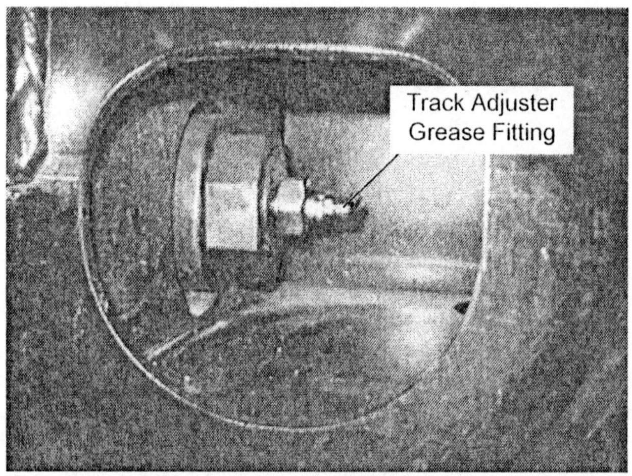

Fig. 17 — Track Adjuster Grease Fitting

Use a standard grease gun to add grease to the track adjusting cylinder grease fitting (1).

Adjusting Tracks

Track performance depends on two vital adjustments:

- Track tension
- Track alignment

If the track is too tight, too loose, or misaligned, it will wear rapidly.

TRACK TENSION

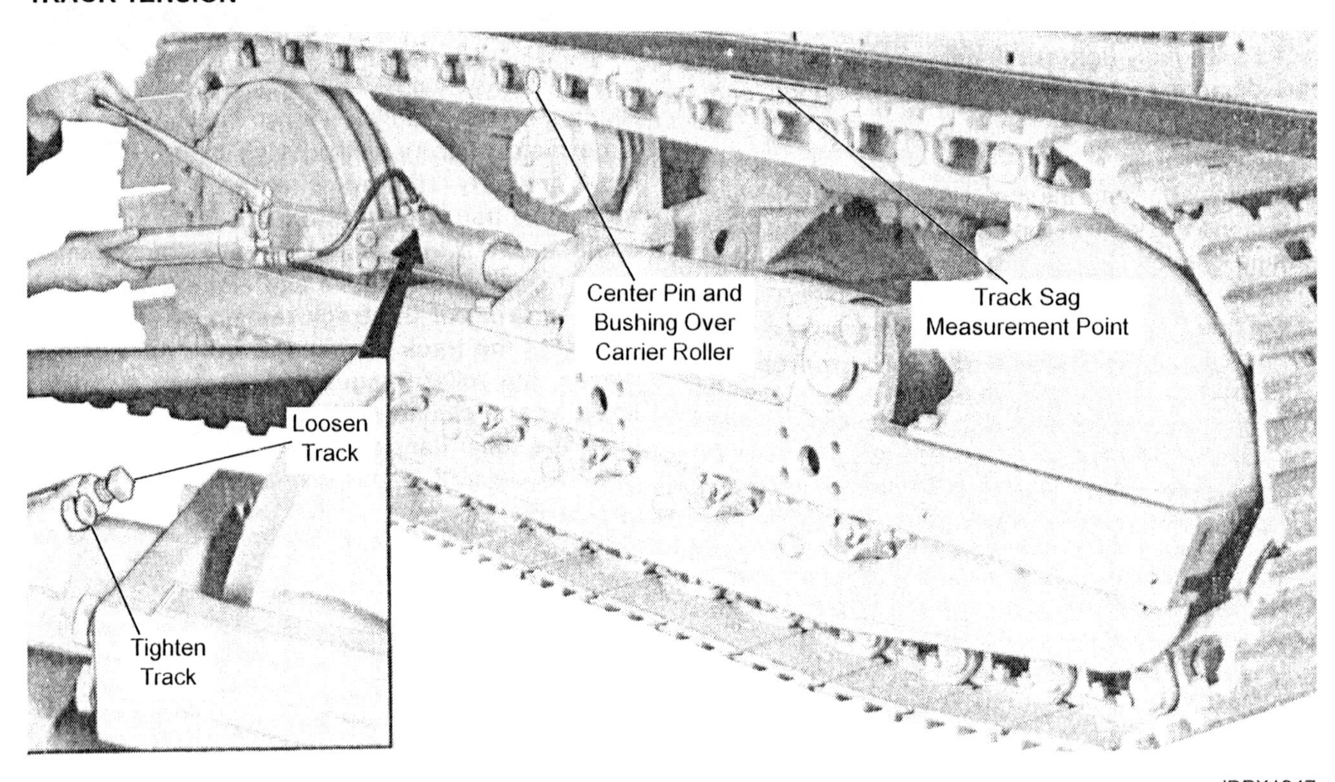

Fig. 18 — Adjusting Track Tension (Hydraulic Mechanism Shown)

For maximum service life of all track components, keep the track properly adjusted. When operating a machine in snow, mud, clay or some types of sandy loam which tend to pack in the track, check the adjustment often. It may be necessary to loosen the adjustment to keep from operating with overtight tracks.

Be sure to check the coils of exposed recoil springs. Mud packing in the recoil spring can cause restriction of the idler travel and recoil action. This can result in overstressing the track links, recoil mechanism and final drives.

Track tension is usually correct if the track has about 1-1/2 to 2 inches (4-5 cm) of sag, measured at a point halfway between the carrier roller and the front idler (Fig. 18). (See the machine operator's manual for specific recommendations.)

To check track tension, measure the amount of sag with a straightedge. Be sure to center a pin and bushing over the carrier roller, if recommended. (If the machine has no carrier roller, center a pin and bushing over the front idler.)

Adjust the tension by either turning the mechanical adjuster or by applying or releasing grease on hydraulic adjusters as shown in Fig. 18.

On hydraulic adjusters, tighten the track by applying grease with the special gun. (Be sure the gun doesn't have too much pressure for the mechanism.) If the piston cannot be moved, disassemble the tension mechanism and free the seized parts.

To loosen the track, turn the relief valve out slightly to relieve hydraulic pressure on the track.

CAUTION: The grease for the track adjuster cylinder is under high pressure. Never remove the special fitting to release the grease. If the grease does not escape from the vent hole with the relief valve open, slowly drive the machine forward and reverse until the grease escapes from the vent hole.

After adding or removing grease, operate the machine to allow the track adjuster cylinder to adjust and then check for proper sag again. Be sure to adjust each track for equal tension.

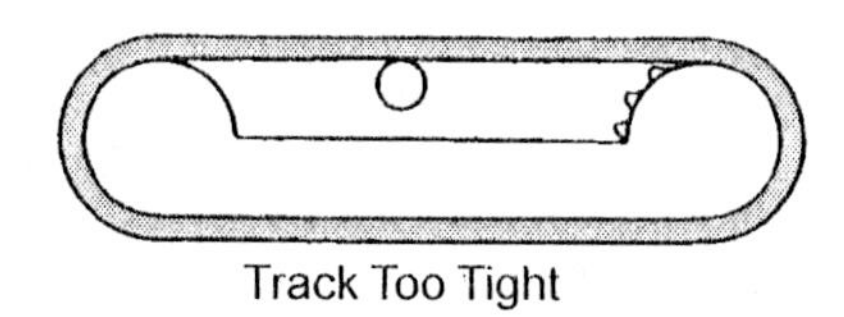

Track Too Tight

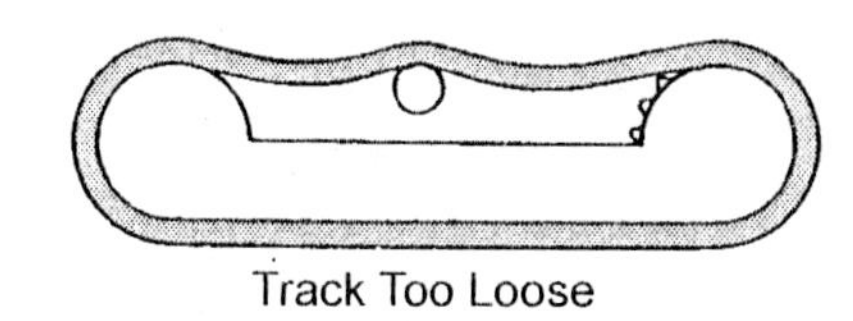

Track Too Loose

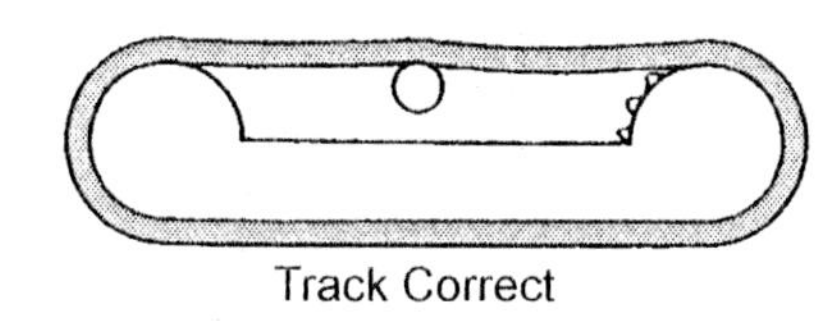

Track Correct

JDPX1348

Fig. 19 — Track Tension

Tracks Too Tight — The Results

If a track is adjusted too tight, a great amount of friction will exist between the pins and bushings as the track links hinge traveling around the sprocket and idler. This friction causes accelerated wear to the pins, bushings, links, sprocket, and idler. Friction from a tight track also robs the tractor of horsepower.

Severely tight track adjustment can cause the track to run extremely hot and "draw-back" the hardness of pins and bushings. In some extreme cases, the pins and bushings may become so hot that they fuse together.

Also, extremely tight track adjustment can cause severe damage to the final drive hubs, bearings, and gears.

The track tension around the sprocket increases dramatically when the track is run too tight. Actual field tests to determine the effects of a tight track on external bushing life revealed that when using 1/2 inch (1.25 cm) track sag, the bushings were due to be turned at about 1100 hours, compared to 2800 hours on a properly adjusted track used in the same test.

Tracks Too Loose — The Results

If the track adjustment is too loose, service life is similarly reduced.

A loose track fails to stay aligned properly and tends to come off when the tractor is turned, causing wear to the idler center flanges, roller flanges and the sides of the sprocket teeth. A loose track will "whip" at high speed, resulting in impact loads on the carrier rollers and their support brackets.

A loose track tension adjustment also permits the sprocket to jump teeth easily, especially in reverse, causing unnecessary wear to sprocket teeth and track bushings.

Another serious result of a loose track, particularly on large tractors with long track, is the failure of the track to stay properly aligned with the rollers. This is especially true if the tractor is operating in rocks or subjected to side loading from hills. For example, a tractor operating in rocks and slopes can have the front and rear of the track resting on rock and the center of the track suspended free. As the tractor moves, the roller flanges will ride up on the track links, wear a chamfer on the edges of the links, and wear the roller flanges.

TRACK ALIGNMENT

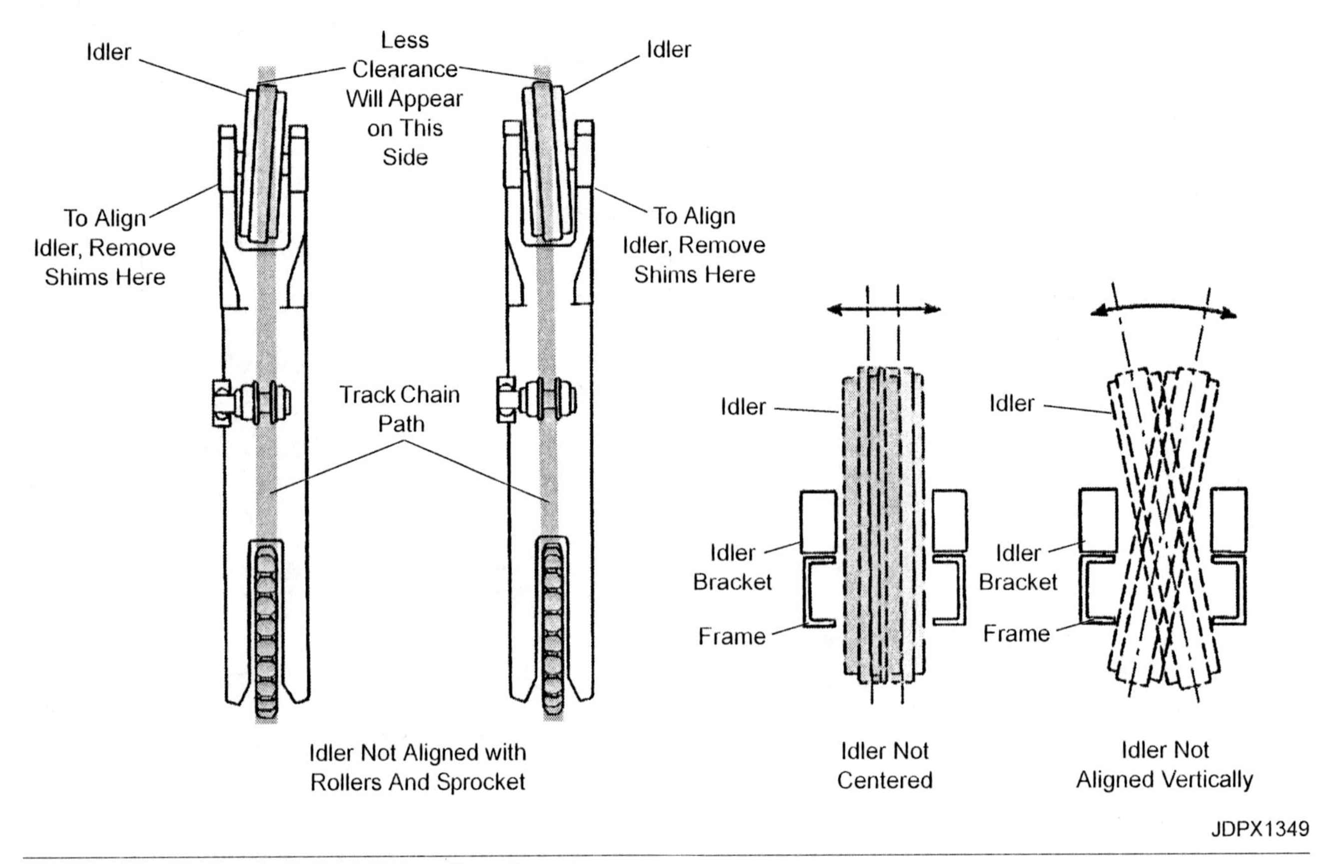

Fig. 20 — Idler Misaligned with Track Parts — Three Examples

If the sides of the drive sprocket and front idler flanges show heavy wear, the track is probably out of alignment.

The sprocket, rollers, and idler should all be in alignment. The frame and rollers are usually aligned first with the sprocket. Then the idler is aligned with the rollers and sprocket.

Most misalignment occurs at the idler. In Fig. 20 you can see typical examples of idler misalignment. When the idler is not aligned with the rollers and sprocket, the idler ridge, the track links, and the rollers will wear rapidly. If the idler is not centered, the wear will be on one side.

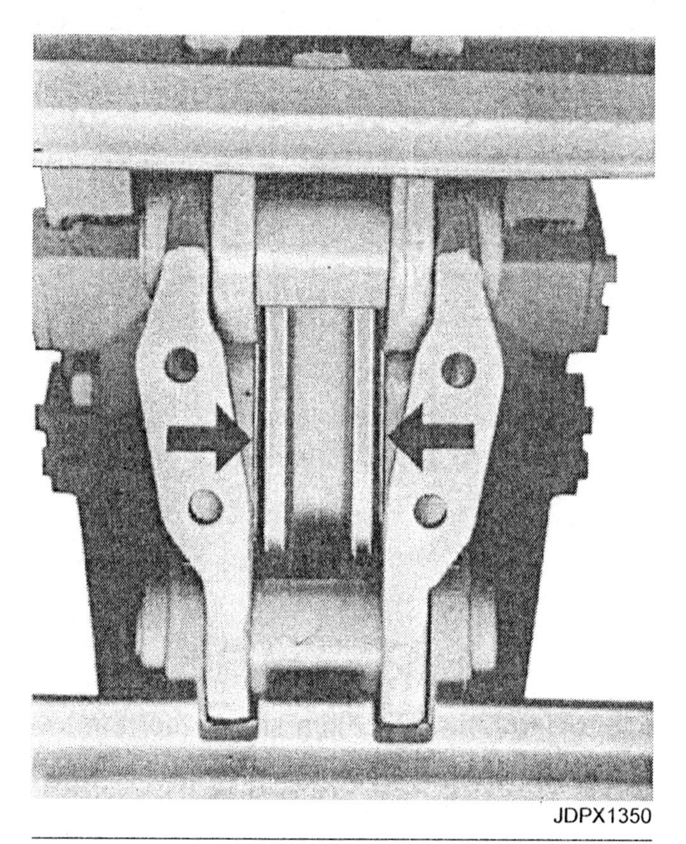

JDPX1350

Fig. 21 — Checking for Equal Clearance at Track Front Idler

If wear is evident, check for the cause. To determine if the idler is aligned with the rollers and sprocket, drive the machine forward on level ground for about 100 feet (30 m). Stop the machine without touching the steering levers. Examine the position of the track links in relation to the front idler flanges (Fig. 21). If the clearance is not equal on both sides, adjustment is required. (It may be necessary to remove a shoe as shown, to check the clearance.)

Notice in Fig. 21 the path of the track chain when the idler is not aligned with the rollers and sprocket.

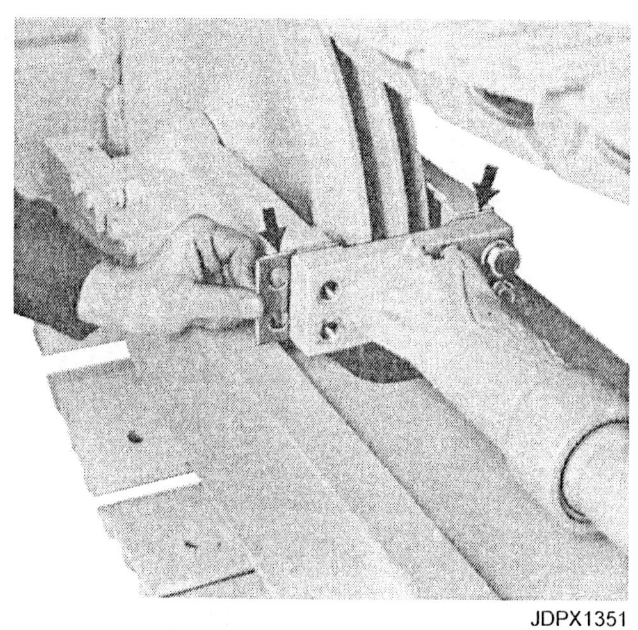

JDPX1351

Fig. 22 — Using Shims to Align Idler with Rollers and Sprocket

If the idler is worn on both sides, it is not aligned with the rollers and sprocket. To correct it, remove the idler adjusting shims from the side of the idler that has the least clearance (Fig. 22). Check alignment again after driving forward another 100 feet (30 m). If the correct alignment cannot be achieved by removing shims, check for a bent frame.

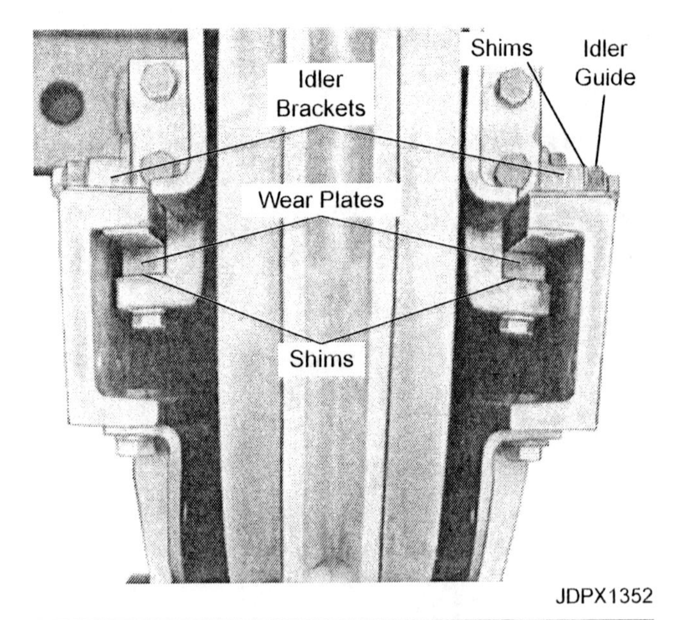

Fig. 23 — Using Shims to Align and Center the Idler

If the idler is worn on one side, it is not centered or aligned with the rollers. To correct it, remove or add shims from each side of the idler bracket (Fig. 23).

To correct vertical misalignment, remove or add shims on each side of the idler. These shims are usually between the underside of the frame and the idler bracket, and between the topside of the frame and the idler bracket (Fig. 23).

Another type of alignment is having the tracks aligned with each other and the machine. This is referred to as toe-in or toe-out. Some manufacturers do not recommend any toe-in or toe-out. If no recommendation is given, both tracks should be parallel to each other and the machine. Also, remember that all the misalignments discussed in this section can be present at once.

If you have difficulty aligning any components, check the track frame for twists or bends. When the frame is twisted, it can often be straightened by changing its position. Do this by loosening and remounting it on the machine or by adding or removing shims from attaching points.

RESULTS OF IMPROPER TRACK ADJUSTMENT

Track Too Loose

- Extremely fast wear on pins, bushings, and track links.
- Unnecessary and rapid wear on sides of drive sprocket, teeth, and idler wheel flanges.
- Possible damaged or broken drive sprocket, idler and idler bracket, side frames, and rollers.
- Track may jump sprocket in both forward and reverse operation. Track may be thrown off when tractor is turned.
- Noisy track.
- Frequent accumulation of trash in track.

Track Too Tight

- Loss of drawbar power and speed. Tractor will not handle rated working load.
- Tractor drifts right or left.
- Fast wear on pins, bushings, and track links.
- Rapid wear on drive sprockets and idler wheels. Extra strain on entire track system because flexibility is lost.
- Unnecessary wear on final drive bearings and oil seals.
- Abnormal steering clutch wear.

Track Misaligned

- Machine drifts away from a direct course.
- Abnormal wear on idler wheel flanges and front idler flanges.
- Excessive track, link, and drive sprocket wear.
- Rapid steering, clutch, and brake wear.
- Operator annoyance and fatigue caused by constantly steering the machine to keep it from drifting.

Diagnosis of Track Wear

A track operated in abrasive sand or rocky conditions cannot be expected to give the service life of one used in soft dirt or snow.

Track components are built tough, but they also get hard wear when large crawlers are operated at high speeds.

Understanding track wear and how to remedy it can be the biggest factor in an effective maintenance program for crawler tractors.

DRIVE SPROCKET WEAR

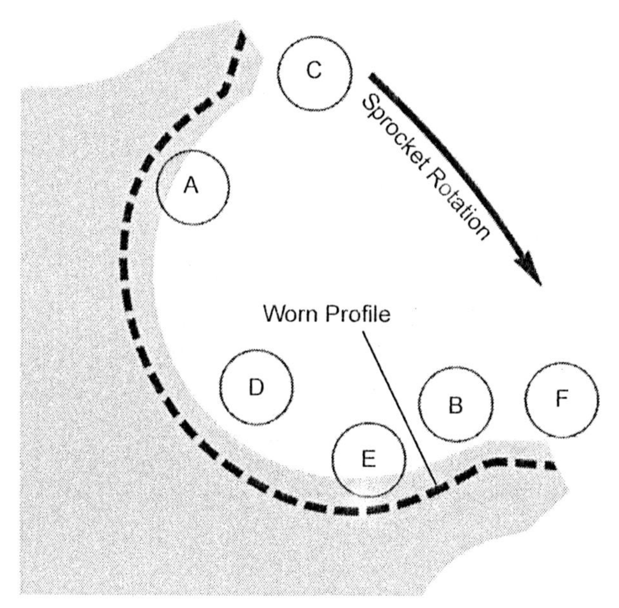

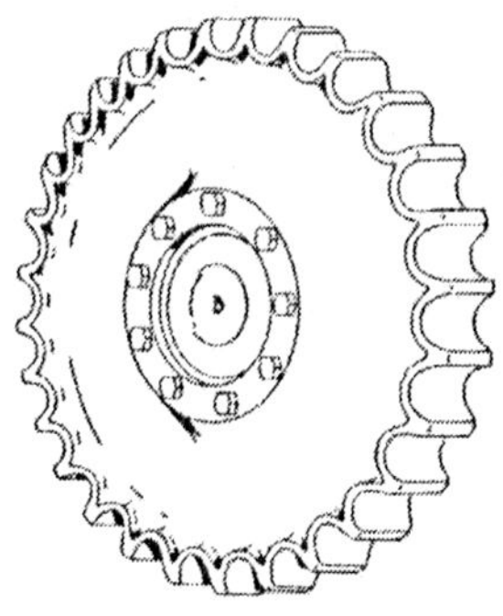

JDPX1353

Fig. 24 — Six Major Types of Sprocket Wear

Sprocket wear is affected by loads, types of terrain, and the abrasiveness and moisture content of soil.

Wear on the sprocket teeth can be identified by careful inspection for six major types of wear (Fig. 24).

- Drive Side Wear (A) — when operating forward.
- Reverse Drive Side Wear (B) — when operating in reverse.
- Climbing Wear (C) — result of increased track pitch.
- Root Wear (D) — when bushing slides from side to side.
- Rotating Wear (E) — when bushing rotates as it leaves (forward) or enters (reverse) the sprocket.
- Reverse Drive Side Tip Wear (F) — when sprocket pitch is greater than track pitch.

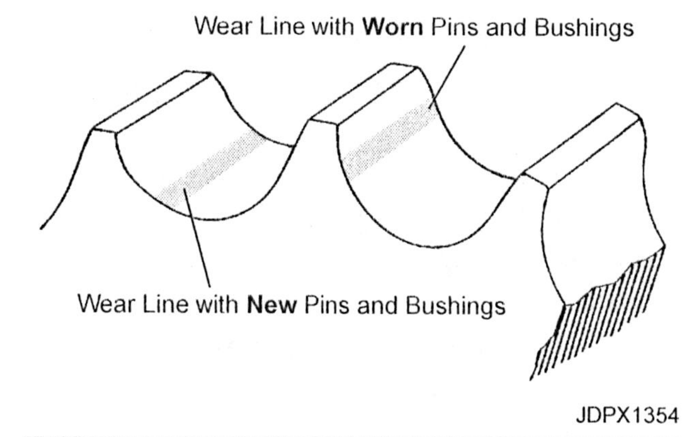

JDPX1354

Fig. 25 — Wear Lines on Sprocket Teeth

Wear lines on a new sprocket are shown in Fig. 25. With worn pins and bushings, the wear line is higher on the tooth.

PIN AND BUSHING WEAR

The master pin and bushing on a conventional track wear more rapidly than the other pins and bushings because there is less contact area between the master pin and bushing, and because sometimes the master bushing does not extend into the counterbores of the master links.

Because of this more rapid wear, the pins and bushings on either side of the master pin for a distance equal to half the circumference of the sprocket usually exhibit more wear than the remaining pins and bushings.

Experience shows that periodic replacement of the master pin reduces the wear on the adjacent pins and bushings. This in turn reduces the sprocket tooth wear.

The split master link used in some chain assemblies utilizes pins and bushings that are identical to the pins and bushings used in the other links of the chain. This provides uniform wear life of all the pins and bushings in these chains.

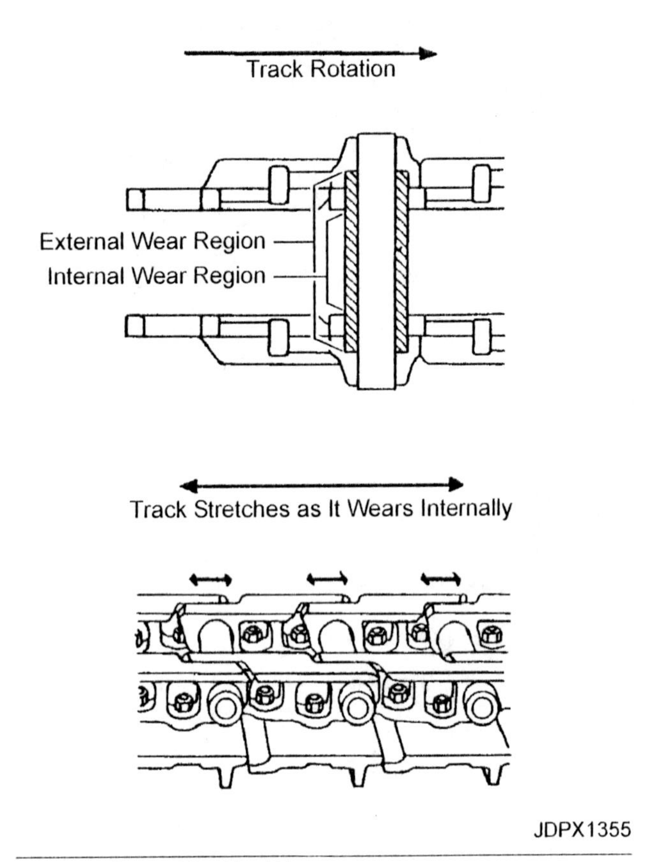

Fig. 26 — Pin and Bushing Wear

There are two types of wear on pins and bushings (Fig. 26):

- External wear
- Internal wear

External wear takes place on the bushings in the area contacted by the sprocket driving teeth, which is about 1/3 of the surface of the track bushing.

Internal wear occurs on the outside diameter of the pin, the inside diameter of the bushing, and the outside diameter of the ends of the bushings if they fit into the track link counterbores (interlocking-type chain). However, it should be noted that internal wear of the pins and bushings is virtually eliminated on a sealed and lubricated track drain.

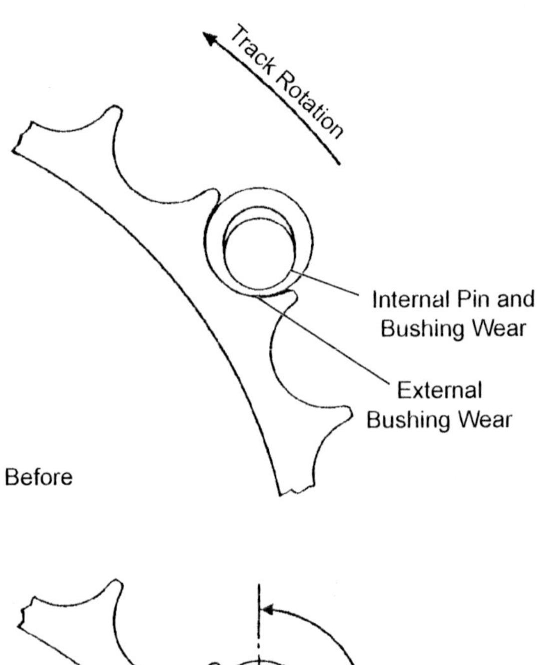

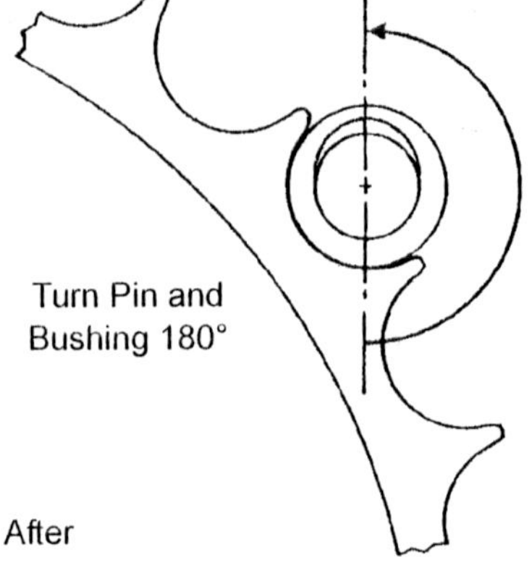

JDPX1356

Fig. 27 — How Worn Pins and Bushings Can Be Rotated to Restore Them

Since wear on the pins and bushings does not extend over the entire surface of the parts, the life of the worn pins and bushings can be increased by rotating them 180 degrees (Fig. 27). However, the pins and bushings must be turned before they are worn past their wear limits or they will not be serviceable.

The amount of pin and bushing internal wear that has occurred can be determined by measuring the track pitch (center-to-center distance between the track pins). Compare your measurement with the manufacturer's wear specification for your machine to determine the percentage of wear remaining.

As the pins and bushings wear, the track becomes longer and track tension adjustment is necessary to compensate for the wear. When the track adjustment limit is reached and the track is still loose, an excessive amount of internal wear has occurred and the pins and bushings should be turned or replaced.

Allowing the pins and bushings to be run to destruction can result in the related wear of all the track components that are affected by operating with a loose track adjustment. Also, the pitch of a track with badly worn pins and bushings is greater than the pitch of the drive sprocket teeth, and may cause sprocket "jumping," resulting in the rapid wear of the sprocket teeth.

However, if the sprocket is also worn, the rotated pins and bushings will continue to wear at a faster rate, unless the sprocket is replaced.

When to Turn Worn Pins and Bushings

There are three signs that signify when the pins and bushings should be turned:

- Wear lines on the drive side of the sprocket teeth.
- External wear on the bushings.
- Internal wear on the pins and bushings (determined by measuring across several of the track links).

Watch all three, but when any one of these points is worn to the manufacturer's wear limits, turn the pins and bushings. To measure the track wear, see "Measuring Track Wear."

NOTE: Refer to the machine Technical Manual for the specified wear limits on track parts.

Pin and Bushing Wear when Operating in Reverse

Driving in reverse, especially at high speed, causes a higher rate of wear on the track pins and bushings than going forward because all the load is placed on one sprocket tooth.

Also, operating the machine with packed material between the sprocket tooth and the bushing will cause wear on the reverse side of the pins and bushings.

ROLLER WEAR

The track rollers under a tractor wear at different rates. Switch the rollers in somewhat the same manner that the tires are rotated on automobiles.

Changing the position of the rollers can distribute the wear and extend the life of the roller group. If you do switch the rollers, change them about midway through their life.

Occasionally, a roller may wear more on one flange than another. If so, the roller can often be switched end-for-end.

The flanges of the front and rear rollers must be in good condition to guide the track properly under the intermediate rollers and the sprocket.

IDLER WEAR

Keep the idlers in proper alignment. A misaligned idler will wear the idler center flange, the sides of the track links, the sides of the roller flanges, and the track roller guards.

If the wear from the misaligned idler is confined to one side, the idler is probably not properly centered (see Fig. 20). There are shims beneath the plates on the idler end bearings to center the idler with respect to the track rollers.

UNDERCARRIAGE WEAR

Link Rail Surface Wear

JDPX8184

Fig. 28 — Link Rail Surface Wear

JDPX8185

Fig. 29 — Link Rail Surface Wear

The rollers carry the weight of the machine on the rail surface of the links. In addition, it is the rail surface which contacts the idler tread as the track chain travels over the idler.

Rolling and sideways sliding contact with the roller and idler tread surfaces will cause the link rail surface to wear. This wear reduces "link height".

Rail wear is a major link wear pattern. It is accelerated by soil abrasiveness, machine weight, and track side-to-side movement.

Link Rail Top Wear Uneven (Scalloped)

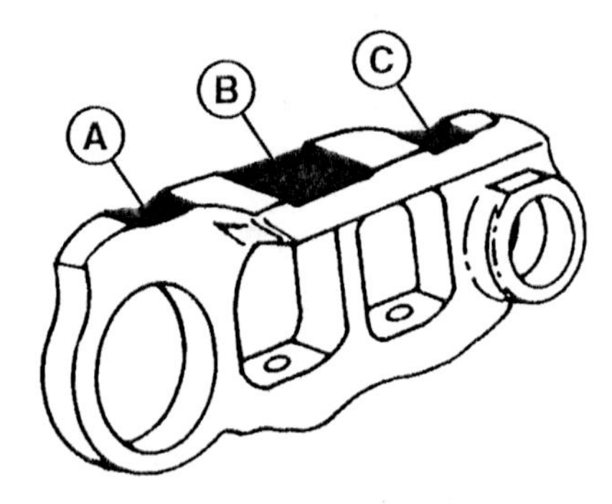

JDPX8186

Fig. 30 — Link Rail Top Wear Uneven

JDPX8185

Fig. 31 — Link Rail Top Wear Uneven

The rail surface may wear unevenly (scalloped wear pattern). The faster wear rate at the center (B) is caused by point contact of the straight link surface on the round idler. Faster wear rate at each end (A and C) is due to abrasion with the roller tread surface when the load from a roller traveling over a joint between link assemblies causes the joint to flex. Scalloped rails result in a rough surface for the rollers to travel on.

Rail wear is a major link wear pattern. It is accelerated by soil abrasiveness, machine weight, and track side-to-side movement. Track sag that is too tight will increase rail wear (A—C).

Link Roll-Over

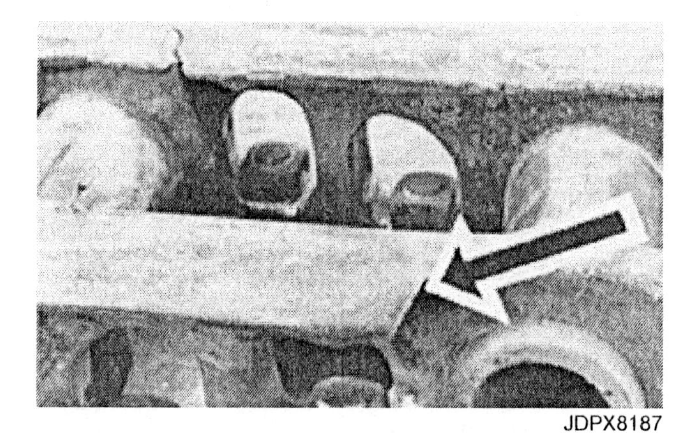

JDPX8187

Fig. 32 — Link Roll-Over

As the rail surface wears, the ends may become peened over by the track roller. This is identified as roll-over.

Roll-over is secondary to rail surface wear and internal wear. It does not affect the function of a track chain. It may, however, be necessary to trim off the roll-over when turning pins and bushings or rebuilding track links.

Link Face Wear

JDPX8188

Fig. 33 — Link Face Wear

Contact between the overlapping faces of track links results in link face wear. This will occur in flush, counterbored, and sealed track chains. There is clearance between these faces in new counterbored or sealed track chains. This clearance is lost and link face wear begins when bushing ends, seals, and counterbore faces become worn. Sealed and lubricated track chain do not experience link face wear since clearance is maintained by the thrust ring.

This type of wear is generally not a factor in deciding the reusability of links.

Rail Side Wear

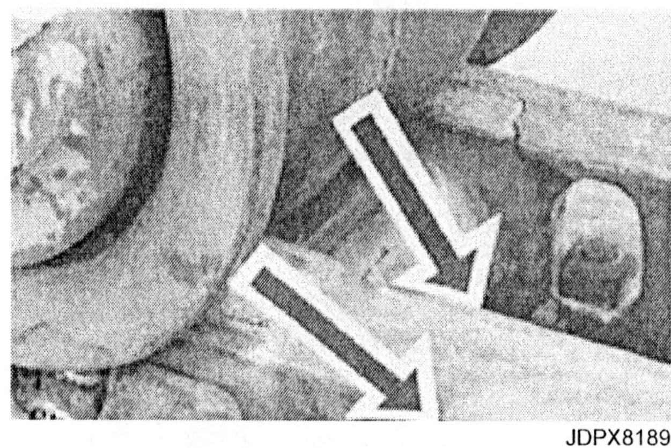

JDPX8189

Fig. 34 — Rail Side Wear

Normal contact with roller and idler flanges results in rail side wear. Side hill operation, turning, and snaky track requires more guiding action and therefore will accelerate rail side wear. Soil abrasiveness is also a contributor to this type of wear.

Pin Boss Wear

Pin boss wear is caused by the boss coming into contact with the roller flange. As rail height and roller diameter decrease, the clearance between the link pin boss and roller flange decreases. Contact and wear on the pin boss will eventually occur when wear is in excess of 100% worn.

Pin boss wear reduces the wall thickness of the boss and the link's ability to maintain a press fit on the pin.

Some pin boss wear is normal when the unit is used on side hills or in applications where a considerable amount of turning is done. This occurs as the pin boss rubs against the track guides. A loose track can also cause the pin boss to come into contact with the track guards when turning or backing up a hill.

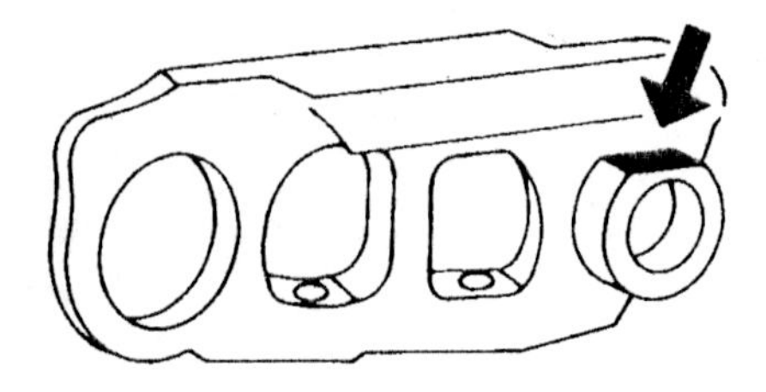

JDPX8190

Fig. 35 — Pin Boss Wear

JDPX8191

Fig. 36 — Pin Boss Wear

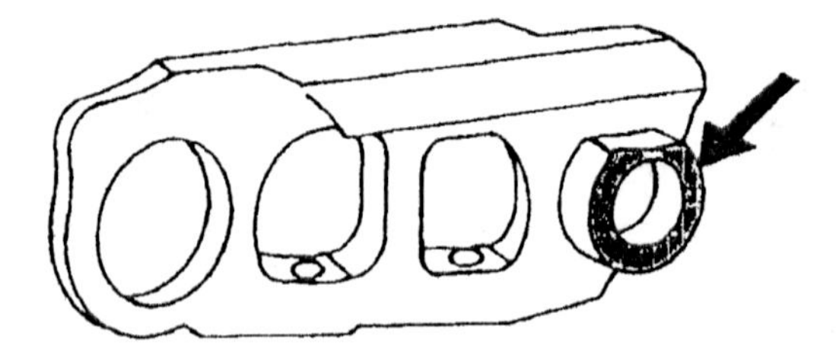

JDPX8192

Fig. 37 — Pin Boss Wear

Link Counterbore Wear

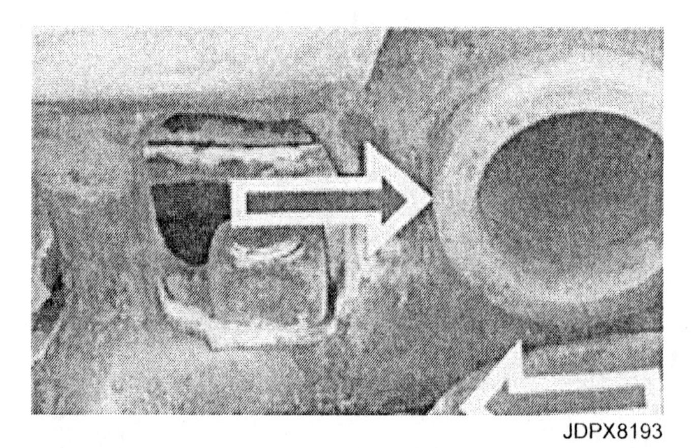

JDPX8193

Fig. 38 — Link Counterbore Wear

In a sealed track, counterbore wear (arrow) is caused by the bushing moving off center due to pin and bushing internal wear. The counterbore in the seal area does not elongate. Sealing ability is not affected as long as pitch is not allowed to go beyond 100% worn.

The sealing washers (C) used in sealed track will wear thin from contact with the end of the bushing and the link counterbore. These are generally replaced when reassembling the track chain and are therefore not evaluated for wear.

Link Cracking

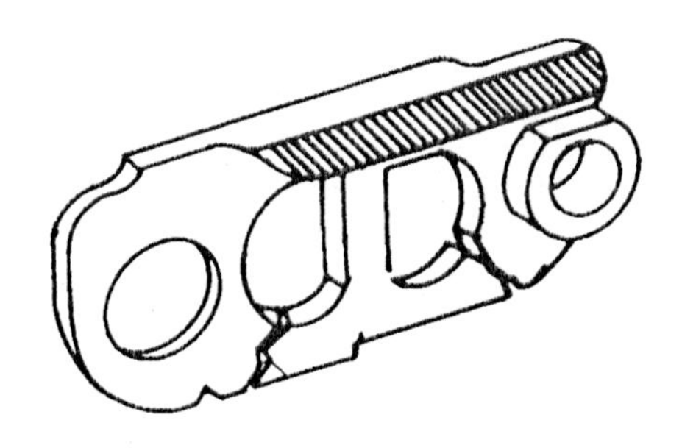

JDPX8194

Fig. 39 — Link Cracking

Track links are subjected to twisting forces as a machine travels over rocks, tree stumps, and other rough terrain. A rock under one end of a track shoe pries up on that end of the shoe and twists its link assembly against neighboring links. The stress concentrates in the corners of link openings and can cause cracks in those areas.

Wider track shoes increase the leverage of a rock or tree stump in twisting the track chain and will therefore contribute to link cracking. Extra machine weight also increases twisting forces and contributes to cracking.

Rail Chipping

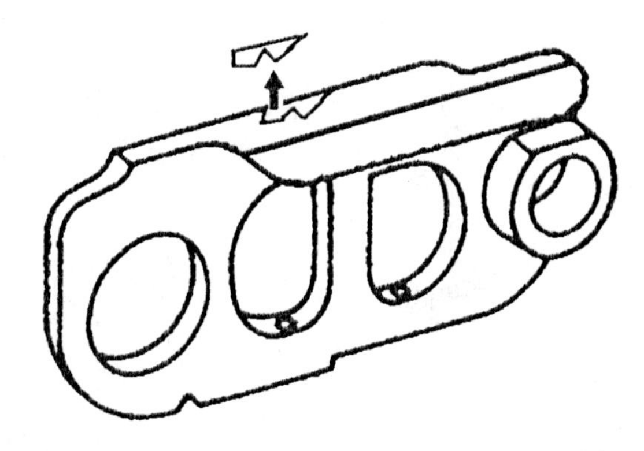

JDPX8195

Fig. 40 — Rail Chipping

Rail chipping is caused by repeated high impacts with roller tread or flanges. This can be caused by a loose or snaky track, or using shoes that are too wide.

Master Pin and Bushing Wear

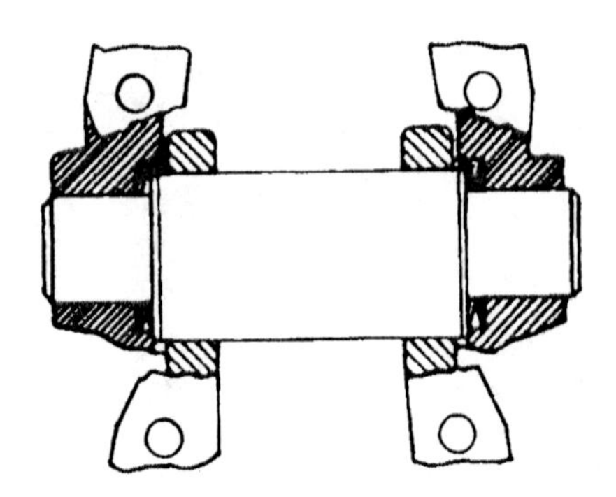

JDPX8196

Fig. 41 — Master Pin and Bushing Wear

The master pin and bushing will wear more rapidly than the other pins and bushings. This is due to the master bushing being shorter than a standard bushing, therefore, providing less contact area with the pin.

Because of this more rapid wear, the master pin should never be included in the section of chain used to measure Track Chain Pitch .

Pin End Wear

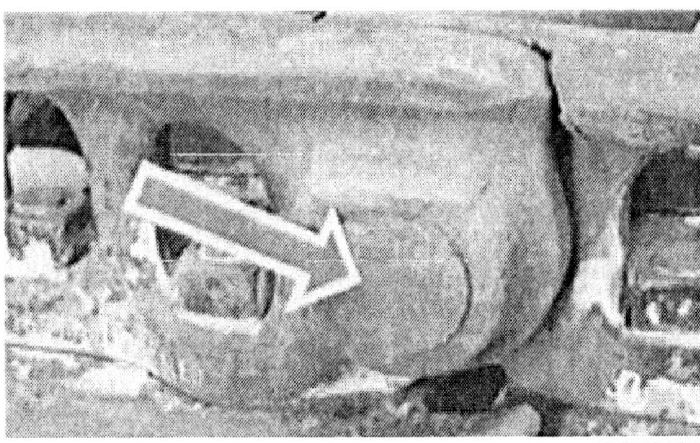

JDPX8197

Fig. 42 — Pin End Wear

Another form of pin wear is that which occurs on the ends. The cause for pin end wear is contact with track guides and rock guides. Side hill operation or turning causes the machine to slide sideways on the track chain. With nearly new rollers and track links, the roller flange will contact the side of the link and prevent pin end contact with guides and guards. As these surfaces wear and as the track chain becomes snaky and more difficult to guide, pin end contact will occur.

Pin end wear is normal and does not affect the usable life of the pin. It is therefore a minor wear pattern.

MEASURING TRACK WEAR

Track wear can be measured in two ways:

- Measuring Tools
- Special Wear Gauges

Measuring tools, such as a ruler, tape measure, depth gauge, and calipers (Figs. 43 and 44), may be used to measure the exact wear of the track components. By comparing the measurements to the manufacturer's specifications, you can tell if the track parts need reconditioning.

Special wear gauges (Fig. 45) are supplied by some track manufacturers to provide a rough guide for determining track wear. However, these gauges do not accurately indicate the track wear. They will tell you if the part is half worn out or if it is worn out beyond the wear limits.

CAUTION: Stop the engine before making any track measurements or adjustments.

Using Measuring Tools

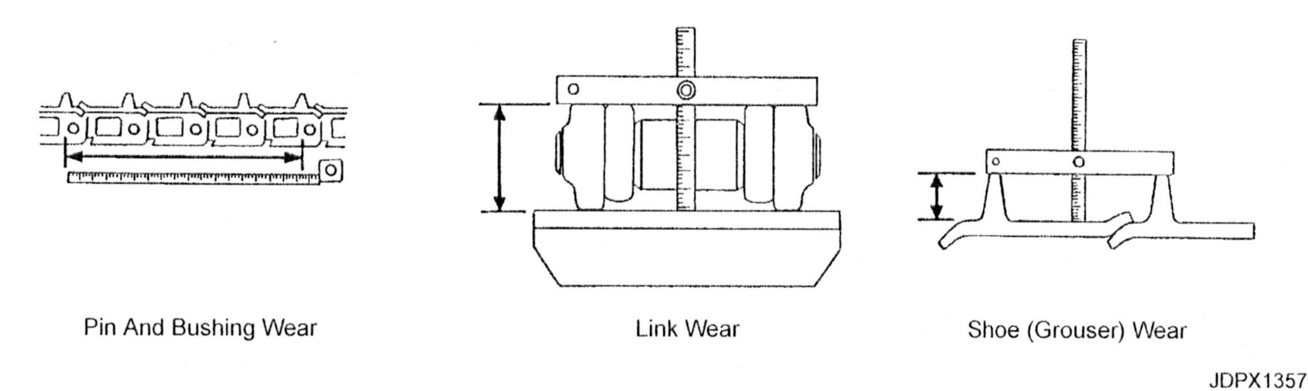

Fig. 43 — Measuring Wear on Pin and Bushing, Link, and Grouser (With Measuring Tools)

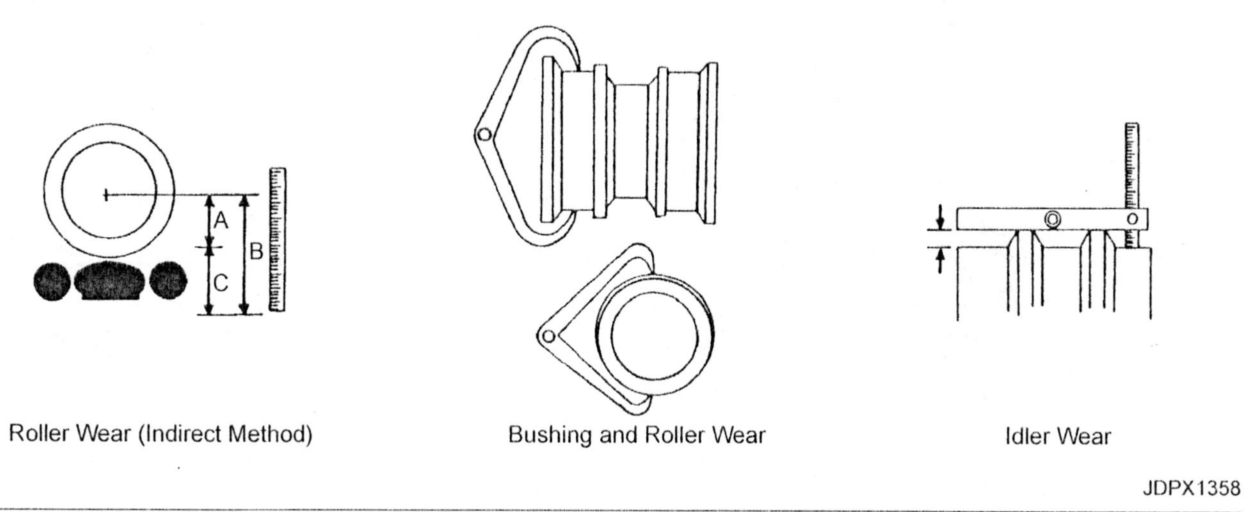

Fig. 44 — Measuring Wear on Rollers, Bushing, and Idler (With Measuring Tools)

Pin and Bushing Wear — The track must be taut to make this measurement (Fig. 43). Place an old track pin between two sprocket teeth and back up until the slack is taken out of the track. Measure from the side of any pin across several links to the same side of another pin. Be sure the master pin is *not* one of these pins. Divide this measurement by the number of links you measured across. This number will tell you how much the pin and bushings have worn internally when you check the figures against the manufacturer's dimensions.

Link Wear — Measure the height of the link with a depth gauge and compare it to the manufacturer's specifications (Fig. 43).

Shoe (Grouser Wear) — Measure the height of the shoe grousers and determine how much they are worn (Fig. 43).

Roller Wear (Indirect Method) — Measure the roller radius with a ruler when the roller guards are installed (Fig. 44). Subtract measurement "C" from measurement "B" to determine roller radius "A."

Bushing and Roller Wear — The track bushing and roller wear can be measured directly with calipers (Fig. 44). Place the caliper ends alongside a ruler and measure the caliper tip spread. Compare the measurement with the wear limit tables.

Idler Wear — Measure the height of the flanges with a depth gauge (Fig. 44). Check the manufacturer's specifications for the recommended wear tolerances.

Using a Wear Gauge

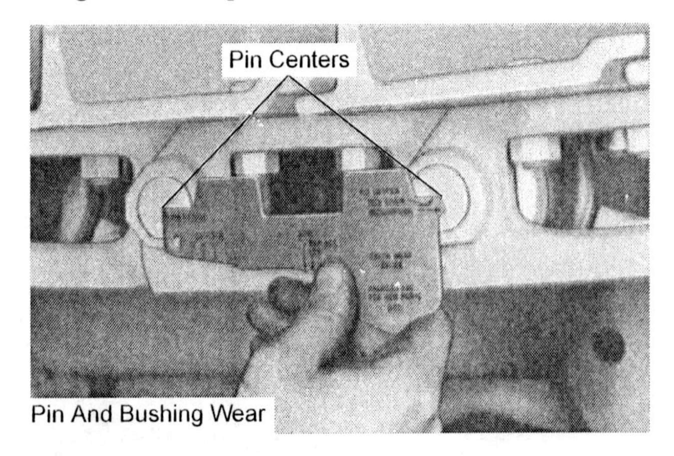

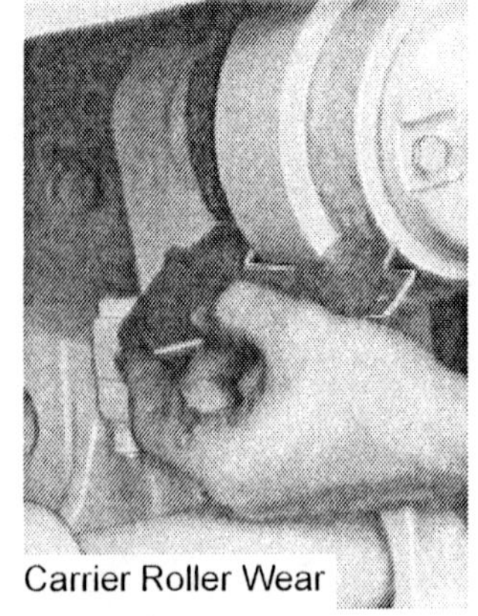

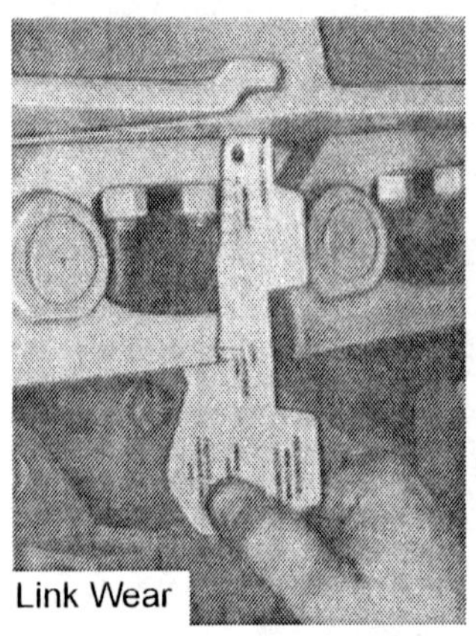

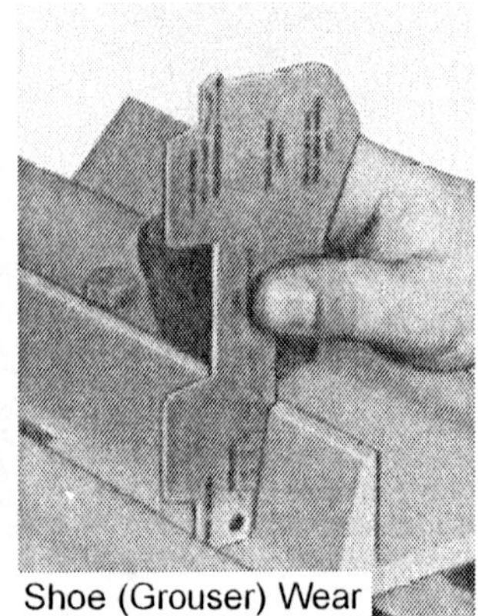

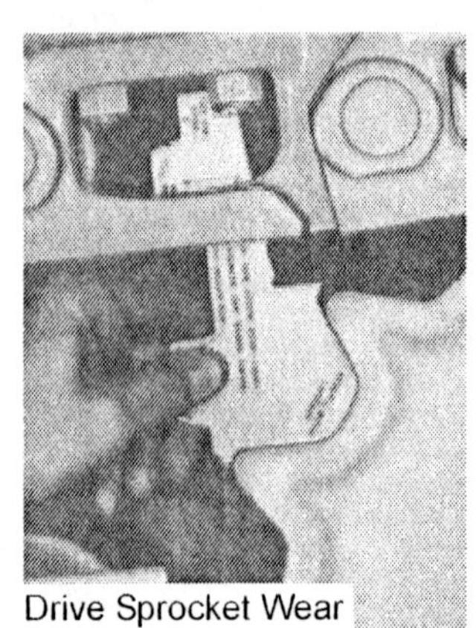

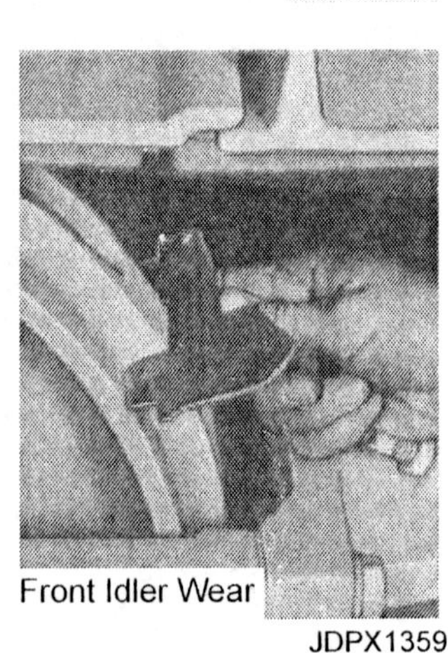

JDPX1359

Fig. 45 — Measuring Track Wear with a Wear Gauge

The instructions below are for a typical wear gauge supplied by one track manufacturer. The shape and markings on the gauge will vary for different tracks.

Pin and Bushing Wear — The track must be taut to make this measurement. Place an old track pin in a sprocket tooth and back up until the slack is taken out of the track. Position the gauge on the track link (Fig. 45). Place the corner of the gauge marked "Pin Center" at the center of one pin and the other end of the gauge at the center of the next pin. If the point of the gauge marked "New Chain" falls at the center of the pin, the track chain does not require servicing. If the point marked "Recondition" falls on the pin center, the track pins and bushings are probably worn.

Link Wear — Position the gauge on the track link (Fig. 45). With the top of the gauge (end with hole) against the track shoe, check the positions of the two arrows on the gauge in relation to the bottom of the link. The link is worn if the arrow marked "Replace" is at the bottom edge of the link.

Grouser Wear — To measure the amount of wear on the grouser lugs, position the gauge upright and against the grouser (Fig. 45). The point on the gauge where the top of the lug falls shows the amount of wear on the grouser track shoe.

Carrier Roller Wear — Place the gauge over the roller support so that the large cutout is over the raise portion of the idler (Fig. 45). The lines on the gauge will match the outer edge of the center flange on a new idler. Wear is indicated by how far the center flange passes the lines on the gauge.

Track Roller Wear — The roller will wear on the sides and the outer surface of the roller flanges. Place the gauge between the roller flanges (Fig. 45). Allowable wear on the outer surface and side of the rollers is shown by the arrows on the gauge.

Sprocket Wear — Place the rounded portion of the gauge against a root of the sprocket as shown (Fig. 45). Compare the wear to the manufacturer's wear limits indicated on the gauge.

Front Idler Wear — Front idlers wear on the sides and outer surfaces. Place the large cutout notch of the gauge over the idler flanges (Fig. 45). Push the gauge to one side against the side of the idler flange. The allowable wear is shown by the arrows on the gauge.

Repairing Tracks — A General Guideline

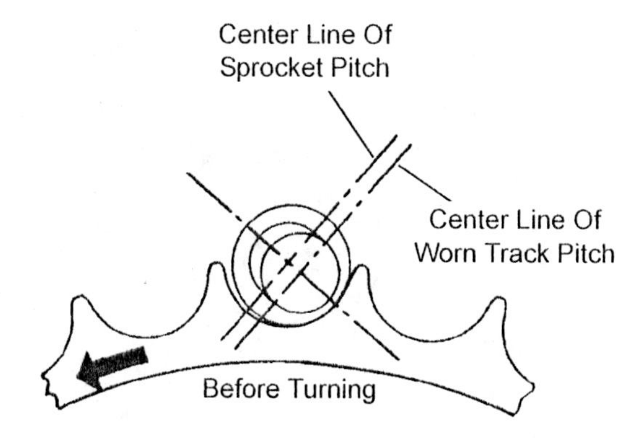

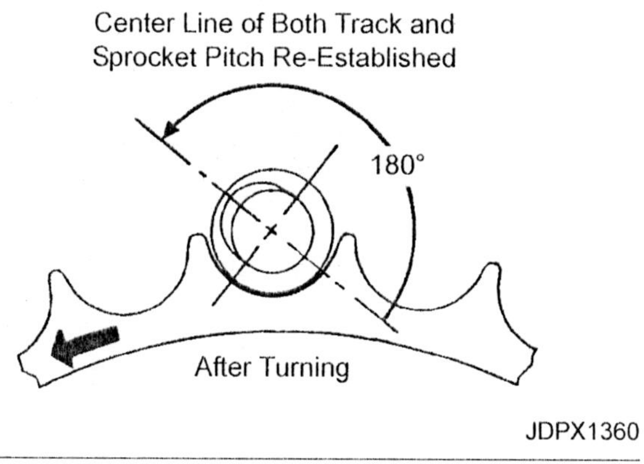

JDPX1360

Fig. 46 — Rotating Worn Pins and Bushings 180 Degrees

The biggest service problem with tracks is pin and bushing wear. Worn pins and bushings can be rotated 180 degrees to renew track pitch, provided they are not worn beyond the limits (Fig. 46).

CAUTION: Do not attempt to repair tracks without the proper equipment and training. Always follow the manufacturer's instructions.

However, it may be more economical to replace the pins and bushings with new parts, depending on the cost of new parts versus the cost of labor.

The pins and bushings in the track assembly are made of extremely hard metal, and are a press fit. The track cannot be disassembled without a track press.

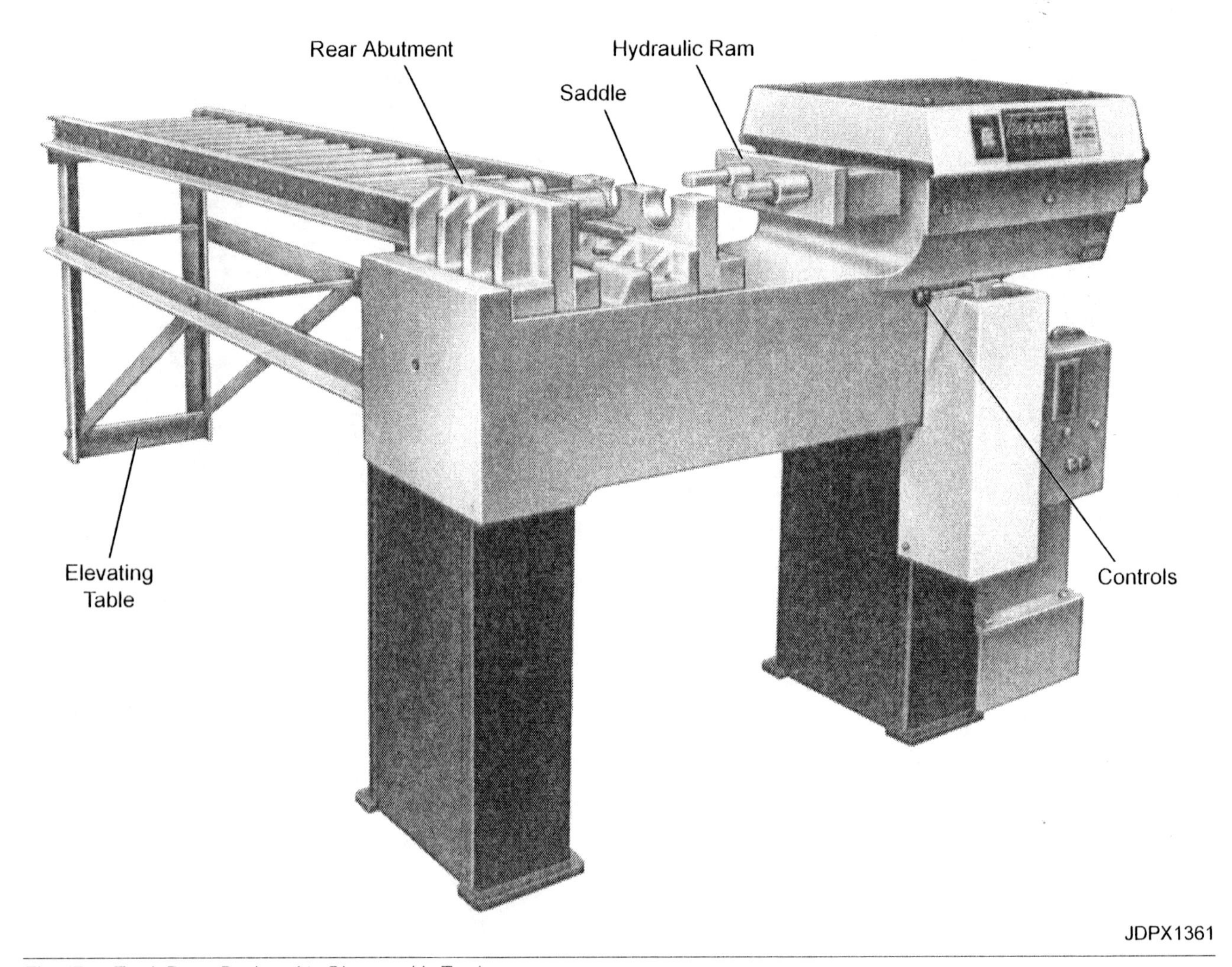

Fig. 47 — Track Press Designed to Disassemble Tracks

Some track presses are stationary presses (shown). Others are trailer-mounted models which may be towed to the job.

Most presses have tools, jigs, and fixtures for all operations in disassembling and reassembling tracks.

Two service methods, with two sets of press adapting tools, are available for track service:

- Interlock-type service
- Flush-type service

The interlock-type service method is used when it is not desirable to push the pin and bushing completely through the link assembly. Both flush-type tracks and interlocking-type tracks may be serviced by the interlock-method adapting tools.

With the flush-type service method, the pin and bushing must be pushed through the link assembly. The adapting tools for this method will accommodate only flush-type tracks.

REMOVING TRACKS FROM MACHINE

Conventional Master Link

To remove the tracks from a machine, perform the following:

1. Raise one side of the machine by placing a floor jack securely under the front crossbar, or an alternate point if recommended, so the track is clear of the floor.
2. Start the engine and shift the transmission into first gear. Pull back on the steering lever that controls the track on the ground.
3. Engage the clutch so the raised track rotates until the master pin is at the 11 o'clock position on the front idler. Note that the master pin is identified by a drill point in the end of the pin.

CAUTION: Do not let the machine move, and be sure the rotating track is clear of the floor.

4. Release all tension from the track to be removed. Place a block of wood under the track near the front idler to support the track in the proper position.

5. Remove the track shoes on each side of the master pin.

6. Remove the snap ring (if used) from the end of the master pin.

CAUTION: Do not hammer on the hardened pins and bushings. They chip.

JDPX1362

Fig. 48 — Removing Tracks from Machine

7. Remove the master pin from the track link using a hydraulic powered master pin pusher/installer as shown in Fig. 48.

8. Operate the machine and rotate the track slowly in the reverse direction to unwrap the track off the drive sprocket.

CAUTION: Keep your hands and feet away from the track during separation.

If the other track is to be removed, repeat these steps.

In the field, the tracks can be removed by separating them at the master pin and driving the machine off the tracks onto planks.

Split Master Link

To remove the tracks from a machine, perform the following:

1. Raise one side of the machine by placing a floor jack securely under the front crossbar, or an alternate lift point if recommended, so the track is clear of the floor.

2. Start the engine and shift the transmission into first gear. Pull back on the steering lever that controls the track on the ground.

3. Engage the clutch to rotate the raised track until the master split link is at the 11 o'clock position on the front idler.

CAUTION: Do not let the machine move, and be sure the rotating track is clear of the floor.

4. Release all tension from the track to be removed. Place a block of wood on top of the recoil spring cover to support the chain at the correct level.

CAUTION: The track chain will fall to the ground when the split link is separated. Do not stand at the end of the track.

5. Remove the shoe at the split link. Separate the split link.

6. Operate the machine and rotate the track slowly in the reverse direction to unwrap the track off the drive sprocket.

TRACK DISASSEMBLY — INTERLOCK METHOD

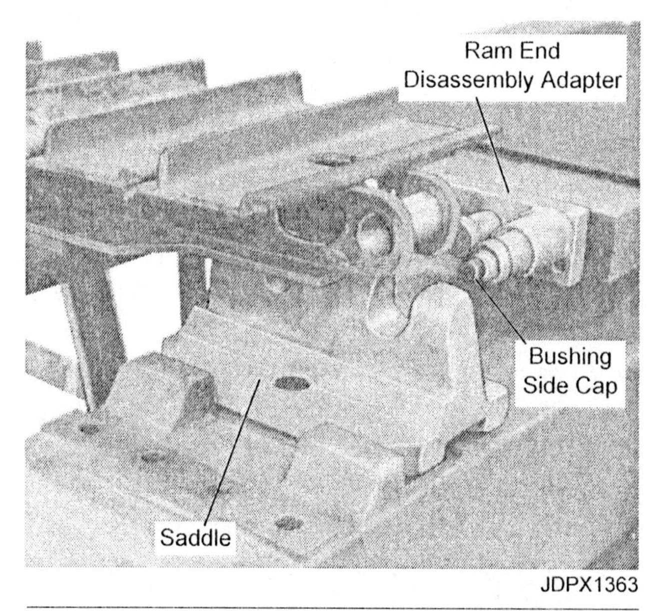

JDPX1363

Fig. 49 — Installing Saddle and Ram End Disassembly Adapter

1. Install the proper saddle and secure it to the track press frame.

2. Adjust the conveyor extension to the desired conveyor working height.

3. Remove the outside row of cap screws, the row away from the work head of the press that secures the track shoes to the link assemblies.

4. Install the ram end disassembly adapter to the work head of the press and secure it (Fig. 49). Install the bushing side cap to the disassembly adapter and secure it.

5. Raise the elevating conveyor and move the track chain assembly toward the press operator until the link assembly bushing is directly over the top of the saddle.

6. Lower conveyor so that the chain link assembly is indexed into the saddle.

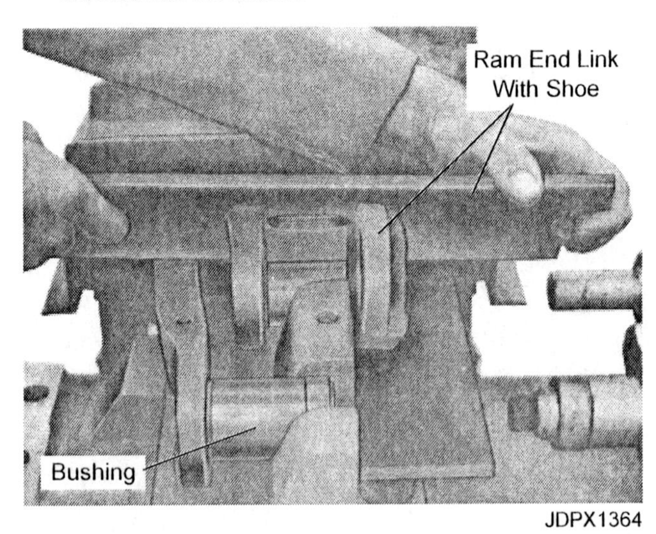

Fig. 50 — Removing Right-Hand Link with Shoe

7. Advance the work head of the track press until the cam and disassembly adapter come in contact with the bushing and pin. Press the bushing and pin from the right-hand link. Remove the right-hand link with shoe as an assembly (Fig. 50).

8. Remove the left-hand link assembly, which includes the bushing and pin.

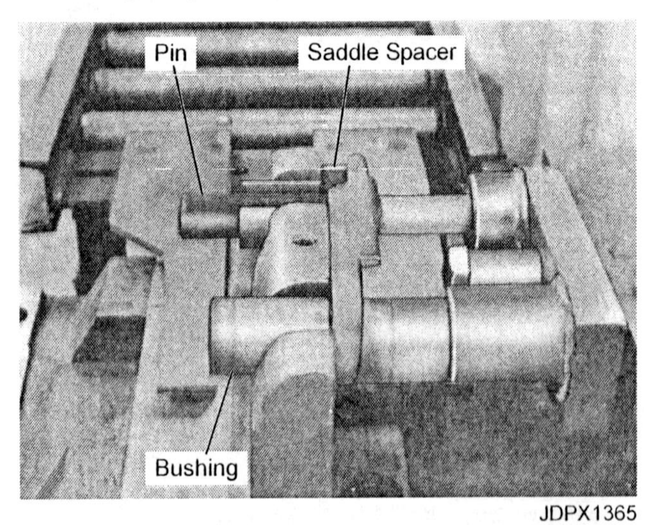

Fig. 51 — Removing Bushing and Pin from Left-Hand Link

9. Install the saddle spacer into the saddle as shown in Fig. 51. Advance the work head of the track press and remove the bushing and pin from the left-hand link.

10. Repeat these steps until the track is completely disassembled.

TRACK ASSEMBLY — INTERLOCK METHOD

1. Remove the cam end disassembly adapter from the press head and install the assembly adapter.

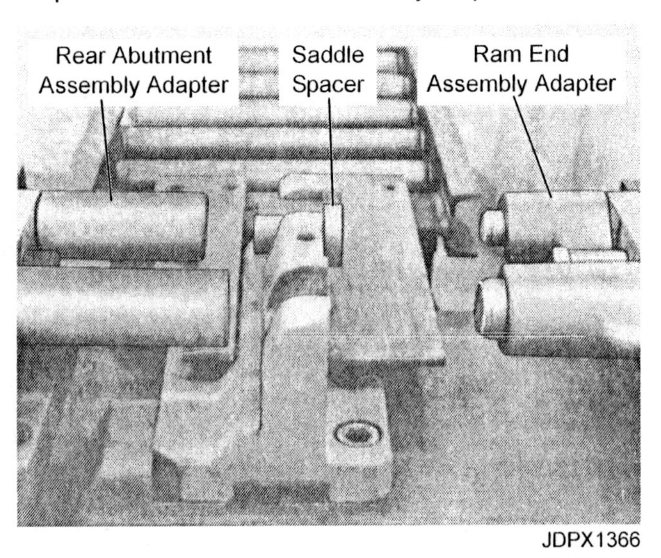

Fig. 52 — Preparing Press for Track Pin and Bushing Installation

2. Install the rear abutment assembly adapter on the press (Fig. 52). Do not remove the saddle or saddle spacer.

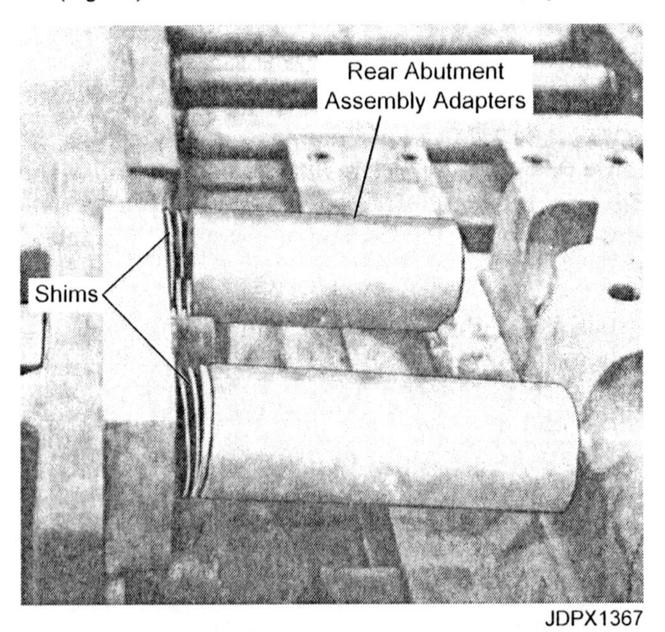

Fig. 53 — Installing Shims behind Abutment Adapter Block

3. Install shims behind the rear abutment assembly adapter blocks to align abutment blocks so that the link assemblies move freely after assembly (Fig. 53). If the link assemblies are pressed too tightly together, the track

will be stiff. As a result, the track will be difficult to install and will wear rapidly.

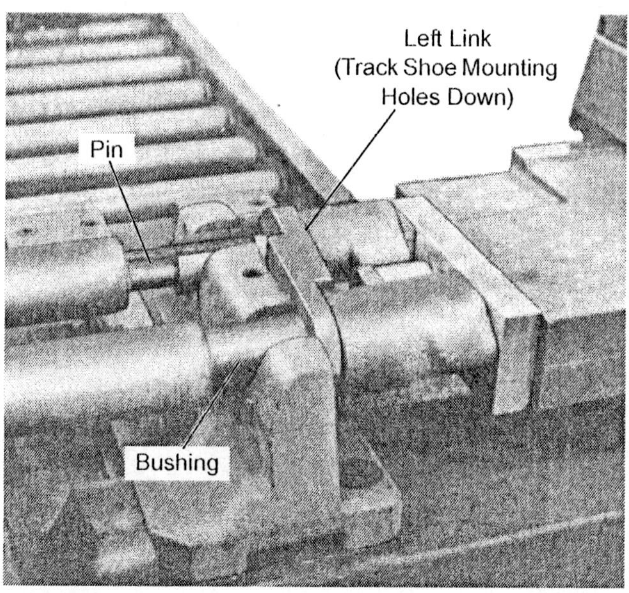

Fig. 54 — Installing Left-Hand Link to Pin and Bushing

4. With the pin and bushing in the saddle, position the left-hand link with the track shoe mounting holes down between the saddle and the work head of the press (Fig. 54). Advance the work head of the press and press the link onto the pin and bushing.

NOTE: *The track shoe mounting holes are in the down position. After the link is pressed onto the pin and bushing, it becomes the left-hand link assembly.*

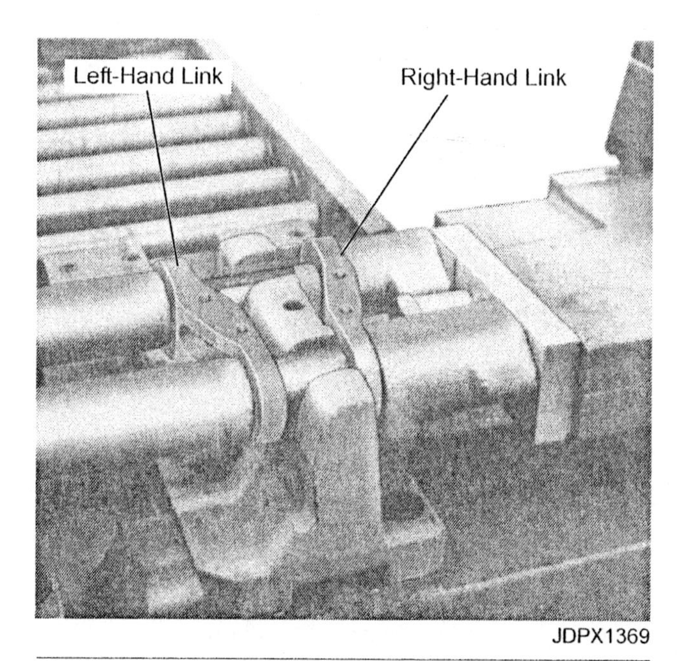

Fig. 55 — Installing Right-Hand Link to the Left-Hand Link

5. Install the right-hand link, with track shoes attached, into proper position (Fig. 55). (Shoe is removed for illustration purposes.)

6. Advance the track pressure work head and press the right-hand link onto left-hand link as shown in Fig. 55.

7. Carefully continue to advance the work head until the right-hand link assembly is properly positioned so the bolt holes in the track shoes align with the bolt holes in the left link.

NOTE: *One link assembly should be completely assembled before assembling the components. This is recommended because it may be necessary, depending upon link wear, to vary the number of adjusting shims behind the rear abutment assembly adapter blocks.*

8. Continue these steps with each link assembly until the track is completely assembled.

TRACK DISASSEMBLY — FLUSH-TYPE METHOD

NOTE: *It is not necessary to remove the track shoes when replacing the pins or bushings unless individual side link replacement is necessary.*

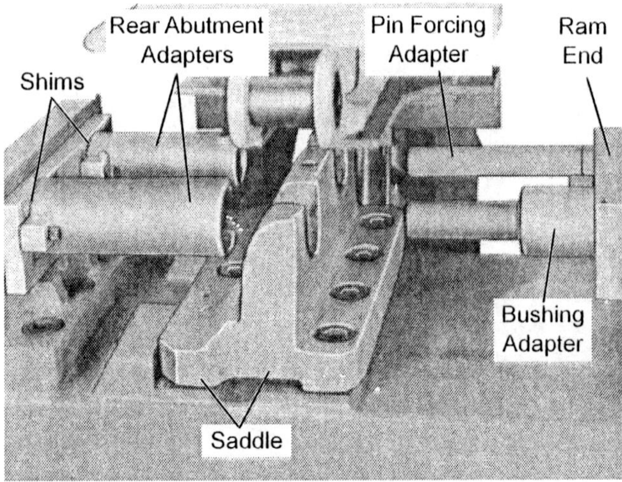

Fig. 56 — Installing Ram End for Disassembly

1. Install the rear abutment adapter with shims (Fig. 56).

NOTE: Due to the friction wear on the side links, the number of shims required will vary. Leave as little space as possible between side link and rear abutment adapter to eliminate any bending stress on the side links when pins and bushings are removed and installed.

2. Install the saddle and secure it to the track press frame.

3. Adjust the conveyor extension to the desired conveyor working height.

4. Install the pin forcing adapter and the bushing adapter on the ram end. Position the ram end on the work head of the press and secure it (Fig. 56).

5. Raise the elevating conveyor and move the track chain assembly toward the operator until the bushing of the link assembly is directly over the top of the saddle.

6. Position the new bushing over the bushing adapter on the ram end.

7. Lower the conveyor so that the chain link assembly is indexed into the saddle.

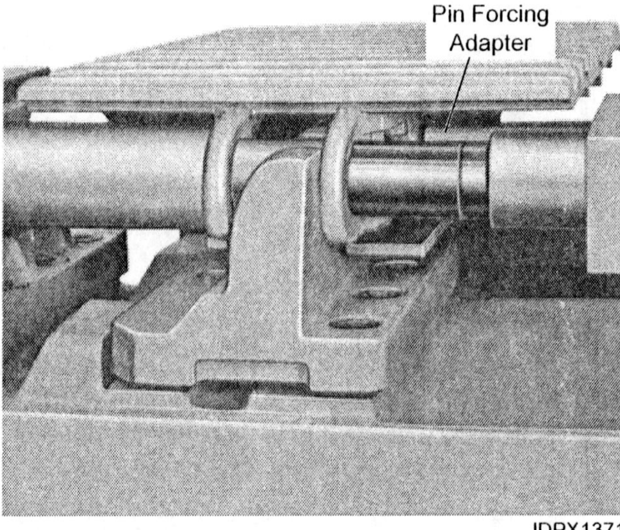

Fig. 57 — Removing Pins and Bushings

8. Advance the work head of the track press until the pin forcing adapter and bushing adapter (with a new bushing) push the pin and old bushing out of the link assembly (Fig. 57).

9. Retract the work head of the track press and raise the elevating conveyor. Remove the link assembly with the new bushing from the track chain assembly.

10. Advance the track assembly and install a new bushing on the bushing adapter. Lower the conveyor so that the chain link assembly is again indexed to the saddle.

11. Keep on removing the pins and forcing out the old bushing with a new bushing until each link assembly is removed from the track assembly.

TRACK ASSEMBLY — FLUSH-TYPE METHOD

1. Remove the ram end from the working head of the press. Remove the pin forcing adapter and bushing adapter from the ram head.

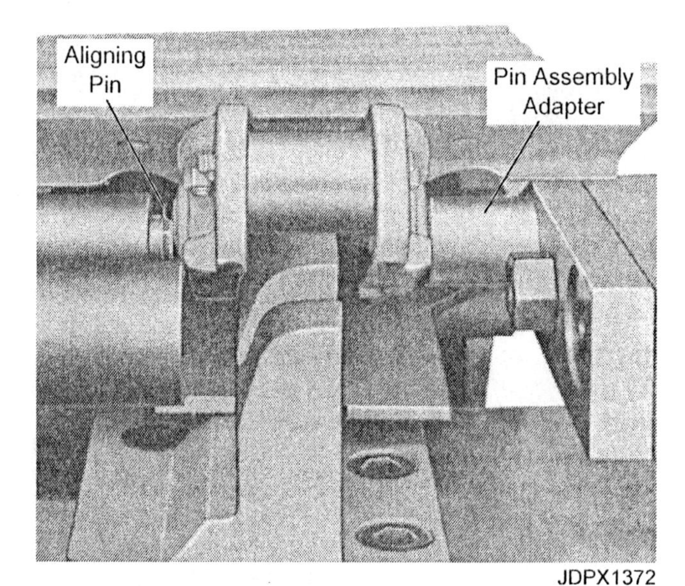

Fig. 58 — Connecting Track Chain

2. Secure the pin assembly adapter to the ram end. Position the ram end on the work head of the press and secure it (Fig. 58).

3. Place one link assembly in position in the saddle so that the bushing rests in the saddle slot nearest the conveyor.

4. Place the second link assembly in position so that the side links straddle the bushing end of the first link assembly. Place the second link bushing in the saddle slot nearest the track press operator. (The link assembly is shown raised in Fig. 58 for illustration only.)

5. Align the link assemblies with the aligning pin. Hold the new track in position and advance the work head of the press. The new pin will push the aligning pin out as the new pin enters the link assembly.

6. Retract the work head of the track press and raise the elevating conveyor to free the link assembly from the saddle. Lower the conveyor and install another link assembly.

7. Continue these steps with each link assembly until the track chain is completely assembled.

TRACK DISASSEMBLY — SEALED AND LUBRICATED CHAIN

See your track press manufacturer's operator's manual for instructions to operate the track press.

Start the disassembly of the chain at the pin-end half of the split link.

1. Check for the proper alignment of the track press with the chain link to prevent broaching of the pin and bushing. The track links cannot be reused if the bores are damaged.

CAUTION: Always wear safety glasses when operating the track press. The hardened parts of the track may break or chip, which could create the risk of personal injury.

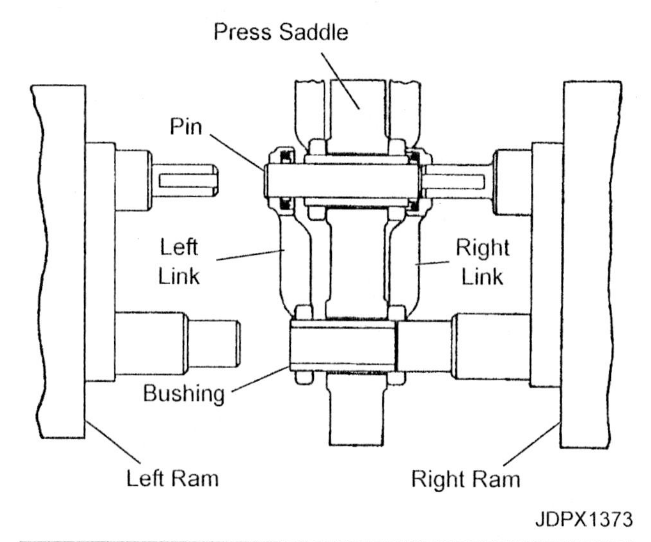

Fig. 59 — Track Press Tools Must Be in Proper Alignment with Links

2. Extend the right ram of the track press to push the pin and bushing from the right link. The right link will be forced against the side of the press saddle as the pin and bushing are pushed out (Fig. 59).

3. Retract the right ram. The right link, with the seals and thrust ring, will stay on the ram tools.

4. Remove the link from the press ram.

5. Extend the left ram to push the pin and bushing from the left link. The link will be forced against the side of the press saddle as the pin and bushing are pushed out.

6. Retract the left ram. The left link, with the seals and thrust ring, will stay on the ram tools.

7. Remove the link from the press ram.

8. Repeat these steps until the track is completely disassembled.

TRACK ASSEMBLY — SEALED AND LUBRICATED CHAIN

1. Apply a mixture of 50 percent alcohol and 50 percent water to the stopper, then install it in the end of the pin bore using a suitable installing tool.

NOTE: The pin end halves of the split master link must be temporarily assembled to the bushing end halves of the link for the proper positioning on the press ram tools. The pin end halves will have to be separated later for the installation at the end of the chain. The split links are a matched set and must be assembled as a set.

2. Assemble the master split link halves using the master shoe bolts. Tighten the bolts just enough to hold the links together. Position the assembled links on the press ram plungers.

3. Adjust the track press pressure relief valve setting to the pressure specified by the track manufacturer.

4. Advance the left ram until the left split link contacts the press saddle. Advance the right ram until it stops to press the split link and bushing assembly together.

5. Measure the distance that the bushing projects from the shoulder of the link using a depth micrometer. The bushing projection determines the clearance between the overlapping links and the proper spacing of the master link bolt holes.

   If the bushing projection does not meet the manufacturer's specification, check for an improper pressure relief valve setting or adjust the thickness of the shim packs behind the ram plungers. Only the first two joints must be checked for proper bushing projection.

6. Apply lubricant to the bushing ends before the next set of links is installed.

7. Install the pin in the bushing and place a thrust ring on each end of the pin.

8. Move the completed split link assembly to the rear seat of the press saddle.

NOTE: The pins must be installed so the cross-drilled hole in the center of the pin is toward the link wear surface or they may break when the chain is used. Install all the pins so the hole in the end of the pins is toward the same side of the chain.

9. Apply a liquid gasket maker to the link bore to prevent the loss of vacuum or lubricant through the pin to the link joint.

Fig. 60 — Installing Load Ring and Seal Ring in Link Counterbore

10. Install the bushing in the front saddle seat (Fig. 60). Install the left link and the right link on the ram plungers. Assemble the load ring and the seal ring, then install the assembly into the link counterbore so the pointed lip of the seal ring is toward the bushing.

11. Advance the left ram until the left link contacts the saddle. Advance the right ram until it stops.

JDPX1375

Fig. 61 — Checking Track Link End Play

12. After one complete joint has been assembled, check the end play of the track links to make sure the bushing, thrust ring and link counterbore faces are pressed solid against each other as follows: Position the base of a dial indicator on one link, and place the indicator pointer against the other link (Fig. 61). Pry the link assemblies in one direction and then in the opposite direction to measure the amount of end play. No (zero) end play should be noted.

    As the track chain is being further assembled, the end play may be checked by flexing each joint after the link assembly is pressed together. The end play is correct if the links cannot be rotated by hand.

    If there is any end play in the chain joint, fabricate a spacer using a section of a bushing. Retract the left ram and install the spacer between the joint in the rear seat of the saddle and the ram. Advance the ram using only the minimum amount of force required to push the joint tight. The pressure should not exceed one-half the relief valve setting.

IMPORTANT: Too much pressure will crush the thrust rings.

13. Add the type of lubricant specified by the manufacturer to the pin and bushing using a suitable seal tester and lubricator tool. Push the lubricator nozzle through the stopper all the way into the pin. Actuate the seal tester to draw a vacuum of 20-30 Hg (68-102 kPa) as indicated on the tester gauge. If there is no decrease in the vacuum for a minimum of five seconds, the joints are sealed. If the vacuum decreases, the joint must be disassembled and the cause of the leak repaired.

14. Add oil to the pin and bushing until the oil pressure (as indicated on the lubricator tool gauge) is 20–30 psi (140–205 kPa). Slowly withdraw the lubricator nozzle from the pin to allow any compressed air in the pin to escape through the second hole in the lubricator nozzle.

15. Repeat the above instructions for all joints.

INSTALLING CONVENTIONAL MASTER LINK TRACKS ON MACHINE

To install tracks, perform the following:

1. Position the track chain under the machine with the wide part of the chain links toward the rear of the machine.

2. Place the end of the chain on the sprocket, then slowly turn the sprocket in the forward direction to pull the chain across the top of the frame to the front idler. Place a block of wood between the upper part of the recoil spring carrier and the track to take the sag out of the chain.

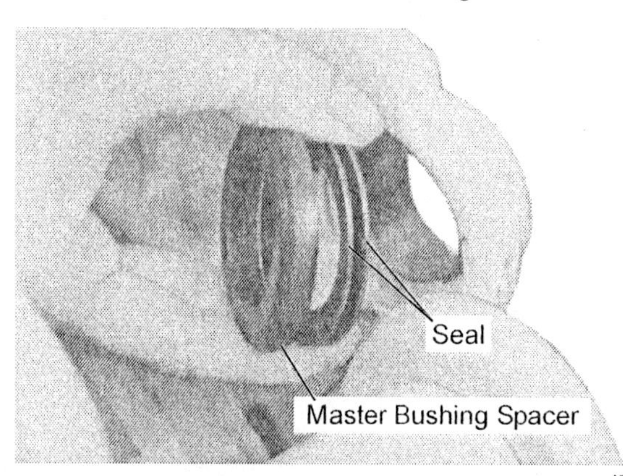

JDPX1376

Fig. 62 — Installing Conventional Sealed Track Chain

3. On a sealed chain, install two seals into the counterbore of the end links, with the inside diameters of the seals touching (Fig. 62). Install the master bushing spacer with the flat side against the seals. Install a forcing pin to align the links, bushing, seals and spacers.

4. Apply oil to the master pin, then use a master pin pusher to press the master pin into the end links.

NOTE: If the master pin is not a tight fit in the links, installation of an oversize master pin is recommended.

5. Lubricate the track shoe mounting cap screws with SAE 30 oil. Tighten the cap screws using a criss-cross sequence to the torque specified by the manufacturer.

6. Adjust the track tension.

INSTALLING SPLIT MASTER LINK TRACKS ON MACHINE

Installing tracks with a split master link is the same as installing conventional tracks except for the following:

1. Locate the split master link at the 11 o'clock position on the front idler. Use a come-a-long device to pull the chain together and engage the split link halves. Apply anti-seize lubricant on the mating surfaces of the split link.

2. Apply anti-seize lubricant on the threads of the shoe mounting cap screws. Install the shoe and tighten the cap screws using a criss-cross sequence to the torque specified by the manufacturer.

Troubleshooting

TRACKS TROUBLESHOOTING CHART

Problem	Possible Causes
Fast wear on pins, bushings, and track links.	Track too loose. Track too tight. Track misaligned. Sprocket badly worn. High-speed operation.
Rapid wear on sides of sprocket teeth, idler, and rollers.	Track misaligned. Sprocket or idler misaligned. Track too loose.
Track whips and damages other components.	Track too loose. Idler frozen in back position.
Track jumps sprocket or is thrown off.	Track too loose.
Noisy track.	Track too loose.
Frequent accumulation of trash in track.	Track too loose.
Loss of drawbar power.	Track too tight.
Drifting to right or left.	One track tighter than the other. Track misaligned.
Abnormal wear on final drive bearings and oil seals.	Track too tight. Track misaligned.
Abnormal wear on steering clutch and brake.	Track too tight. Track misaligned.

Test Yourself

QUESTIONS

1. How do interlocking track chains differ from flush-type chains?

2. Explain "hunting tooth" design of drive sprockets.

3. Why should the grease fitting on the track tension adjuster never be lubricated?

4. What is normal sag for most tracks?

   a. 1/2 inch (13 mm)

   b. 1-1/2 inches (38 mm)

   c. 3 inches (76 mm)

5. What can cause fast wear on track parts?

   a. Too-loose tracks

   b. Too-tight tracks

   c. Both of the above

6. When pins and bushings are worn, will the wear line on the mating sprocket teeth be higher or lower than with new parts?

7. How can worn track pins and bushings be restored?

8. What happens to the length of the track as the pins and bushings wear?

9. How do you measure for pin and bushing wear using a tape measure?

10. What is the function of the master pin in a conventional track? How is it different from the other pins?

11. What is the function of the split master link in a sealed and lubricated track? How is it different from the other links?

12. Match each problem at left below with one of the causes at right.

a. Rapid wear on sides of idlers and rollers.	1. Track too loose.
b. Abnormal wear on final drives.	2. Track misaligned
c. Track jumps sprocket.	3. Track too tight.

ANSWERS

1. Interlocking chains have their bushings extending into the counterbores of the links. This keeps out more abrasives.

2. Only every other tooth engages a track bushing as it rotates. This reduces sprocket wear. (Sprocket has an odd number of teeth.)

3. Because it will overtighten the track.

4. b — 1-1/2 inches (38 mm)

5. c — Both of the above.

6. Higher.

7. By rotating them 180 degrees.

8. The track gets longer, or "stretches."

9. Measure across several links from one pin edge to the same edge of another. Divide the measurement by the number of links and compare to the manufacturer's wear limits.

10. To allow easy removal when separating the track. On a flush-type track, the master pin has a head and is held in place by a snap ring. On an interlocking type track, the master pin is identified by a drill point in the end and is held in place by an interference fit in the link.

11. To allow easy removal when separating the track. The split master link is a two-piece, interlocking link held together by the track shoe mounting bolts.

12. a — 2; b — 3; c — 1.

RUBBER TRACKS

3

Rubber tracks are currently available on today's most powerful tractors as an alternative to pneumatic tires. Rubber tracks provide high tractive capacity, low ground pressure, and high flotation capacity. However, rubber track systems contain significantly fewer components than traditional track systems. This less complex design increases the reliability and reduces the maintenance costs associated with the track system.

A rubber track system does add significantly to the initial cost of the tractor. However, the maintenance cost of a rubber track system is very similar to pneumatic tires.

Rubber tracks are ruggedly built for use in muddy or loose soil. However, reckless operation can put violent shocks on track parts and damage them. Keeping rubber tracks adjusted and maintained correctly will minimize potential damage.

Due to design differences between manufacturers, this chapter should not be considered the final word on rubber tracks. Always follow tractor, track and component manufacturer's handling, operating and repair procedures.

Let's look at the rubber track and undercarriage, how it works and how it can fail.

Operation

Rubber tracks are a large belt with a wide profile, tire-like tread. Each rubber track system has six basic parts:

- Rubber Track — continuous belt
- Drive Wheel — drives the track
- Mid Rollers — support and align track
- Front Idler — maintains track tension
- Tension Mechanism — adjusts track tension
- Track Frame — supports track parts

All these parts together make up the undercarriage. One set for each rubber track, driven by the tractor's final drives, supports and moves the complete machine.

Let's look at the function of each part.

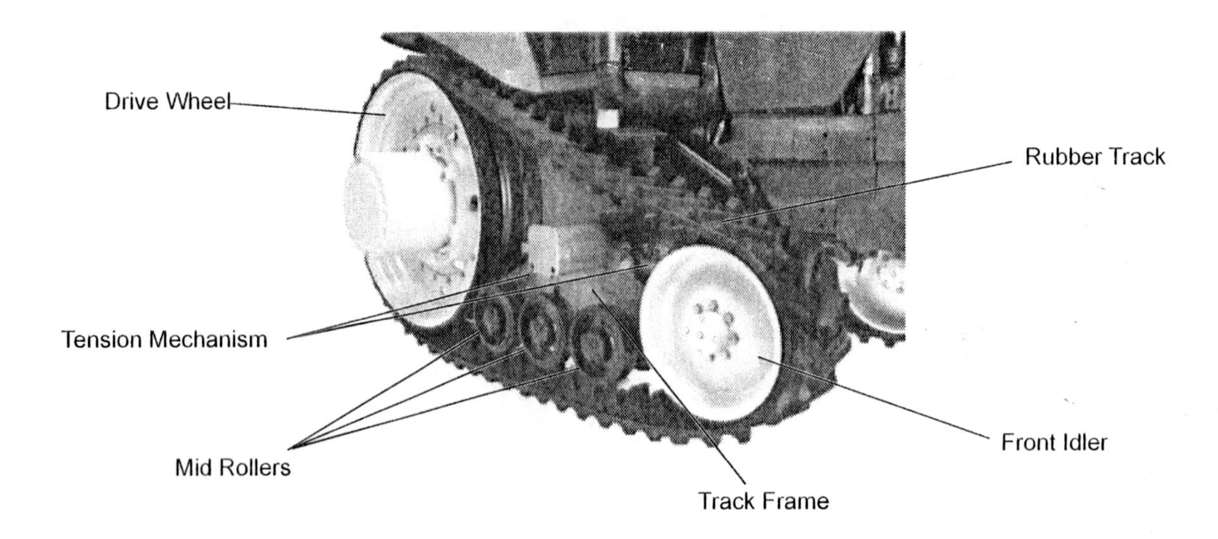

Fig. 1 — Complete Undercarriage of Rubber Track Tractor

RUBBER TRACK

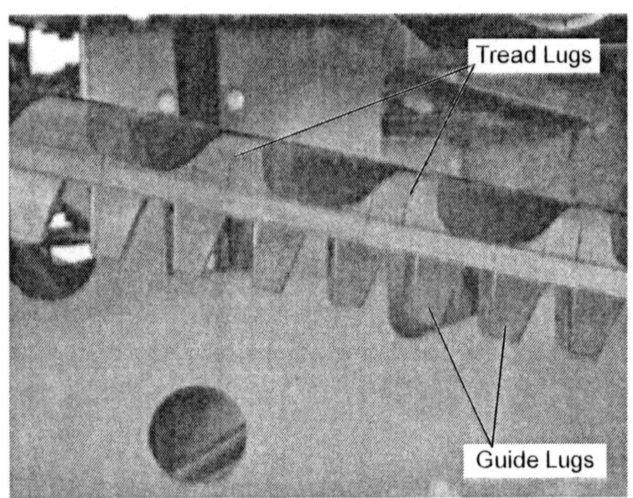

Fig. 2 — Rubber Track Lugs

The track is a continuous belt consisting of a tire-like tread on the exterior side (Fig. 2). Tracks are available in various widths from 16 inches to 36 inches designed to match the intended field application. A single track can weigh up to 1500 lb (680 kg).

To provide strength and stability, the carcass of the track is heavily reinforced with cables and wire plies. A series of guide lugs on the interior of the track maintains alignment as it travels around the drive wheel and front idler (Fig. 2). In positive drive systems, traction lugs are used to assist in power transfer.

DRIVE WHEEL

Fig. 3 — Rubberized Outer Rim of Drive Wheel

The drive wheel is driven by the tractor final drive and transmits power to the track. The outer rim of the drive wheel is rubberized to provide friction for power transfer to the rubber track (Fig. 3).

MID ROLLERS

The mid rollers, mounted on the base of the frame, support much of the machine's weight. They are relatively small and close together to provide even pressure on the track as it revolves and contacts the ground. Typically, there are two rows of rollers, one row on each side of the guide lugs. The tractor weight is distributed evenly over the entire bottom of the track, giving the track its high level of traction and flotation.

FRONT IDLER WHEEL

The front idler wheel supports the track as it revolves. A precharged hydraulic system provides constant, flexible tension on the track through the tension mechanism.

TENSION MECHANISM

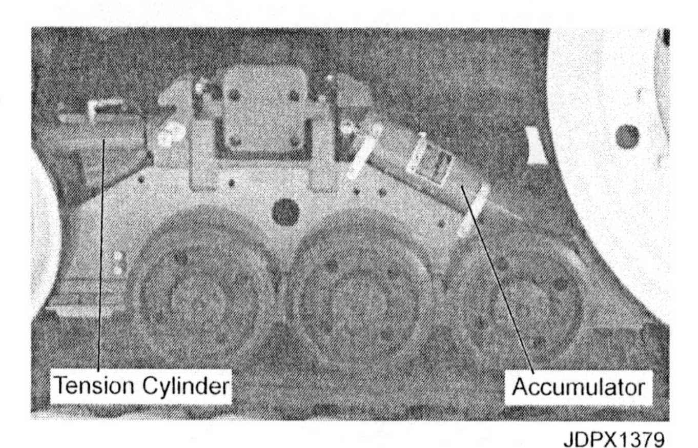

Fig. 4 — Track Tension Cylinder and Accumulator

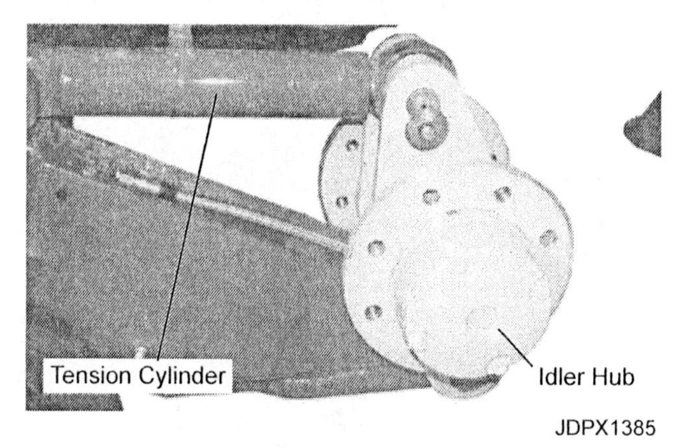

Fig. 5 — Track Tension Cylinder and Pivoting Front Idler Hub

Track tension is maintained with a hydraulic tension mechanism, consisting of a track tension cylinder, charged accumulator, tension sensor, and pivoting front idler wheel hub (Figs. 4 and 5).

This system is actuated hydraulically to tension or detension the track.

TRACK FRAME

The track frame supports the undercarriage components. The front idler wheel hub, mid rollers and the tension mechanism are all attached to the track frame. However, the drive wheel is not supported by the frame. It is supported by the drive axle of the tractor.

The track frame is securely fastened to the tractor mid-frame beam, a section of the tractor chassis.

Cleaning and Inspecting Tracks

CLEANING

Regular cleaning of the undercarriage will reduce wear and increase performance of the track system. Mud and trash build-up not cleaned out may result in increased wear, premature failure, or damage to undercarriage components. Rocks may severely damage undercarriage components if not removed regularly.

INSPECTION

A good **daily inspection** of the track assembly takes only a few minutes but can prevent major problems before they happen.

Inspect the complete track and undercarriage for misaligned, loose, or missing parts. Use a torque wrench to tighten loose hardware.

Rubber Track

- Check for lost lugs, cuts, or chunking on traction lugs. It is normal for guide lugs to show some amount of scuffing between the guide lugs and the drive wheel rubber.
- Check for wear, cuts, or exposed cables in the track carcass. Protruding wires should be cut flush with track.

Wheels and Rollers

- Check for excess wear on the rubber coating, which can be caused by debris collection, low tension or drive wheel slippage in wet conditions. Spaces between rubber and metal could be due to poor adhesion.
- Check for grooving in mid roller rubber coating caused by rocks riding between rollers.
- Check for loss of paint on rims, indicating trash rubbing.

Tension Mechanism

- Check for leaks in the hydraulic system indicated by seepage/dust collection at connections and fittings.
- Check hose for cracks or rubbing.
- Check connections, appearance and protected routing of the tension sensor electrical wiring.

Track Frame and Mid-Frame

- Check frame weld seams and bolted joints for cracks or wear.

Adjusting Tracks

Track performance depends on two vital adjustments:

- Track Tension
- Track Alignment

If the track is too tight, too loose, or misaligned, it could wear rapidly and not perform as anticipated.

TRACK TENSION

Tension adjustment on a track vehicle is critical to ensure proper operation of the tractor and eliminate unnecessary wear on undercarriage components. When tensioning tracks, each side is an independent system and needs to be tensioned individually.

Tractors with rubber tracks monitor the track tension with electronic sensors. Low tension and high tension conditions are communicated to the operator with a warning indicator.

Adjustment should only be made with tractor sitting on flat/ level ground.

Adjusting track tension on a typical tractor is done by attaching a hydraulic hose to a SCV port of the tractor and the track tension cylinder.

Moving the SCV lever in one direction will draw fluid from the tension cylinder until the track goes slack. Moving the SCV lever in the other direction will flow fluid into the tension cylinder to tension the track.

After the track is tensioned, the alignment should be checked and adjusted, if necessary.

TRACK ALIGNMENT

Correct alignment, between the track on the track frame and the guide lugs running in the center of the front idler and mid rollers is a key factor in maximizing the life of the track and frame components.

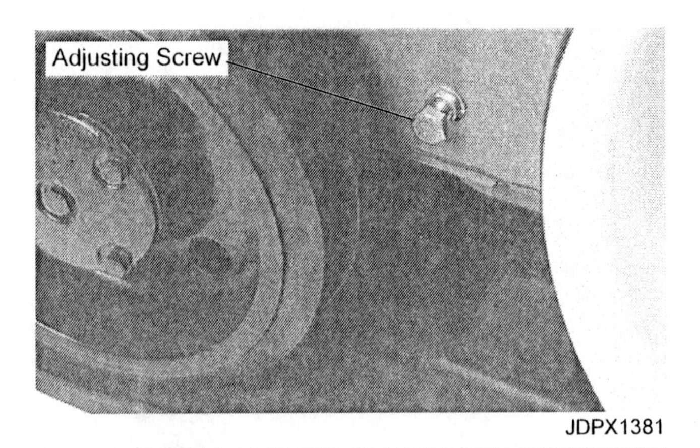

JDPX1381

Fig. 6 — Adjusting Screw

Some manufacturers use adjusting screws, mounted on the track frame, to change track alignment (Fig. 6). Before checking the alignment, make sure both inside and outside adjusting screws are tight on each track frame.

Checking Alignment

To check track alignment:

1. Drive tractor in a straight line for about 1300 ft (400 m).

JDPX1382

Fig. 7 — Temperature Check Position

2. Feel surface temperature of the sides of the guide lugs, about 3/4 in. (20 mm) from the lugs contacting the track (Fig. 7). Little temperature difference between opposite sides of a lug indicates correct alignment. If a higher temperature is felt on one side of the lug, friction is higher on the warmer side of the lug. The alignment of the track should be adjusted.

Adjusting Alignment

To adjust the alignment of the track:

1. Move the track in the direction opposite the hot side of the guide lug by loosening the adjusting screw on the hot side and then tightening the other. Tightening the inside screw will move the track toward the tractor chassis and tightening the outside screw will move the track away from the chassis.

NOTE: Do not adjust the screws by more than 1/2 turn at a time. Retighten lock nuts before driving the tractor.

2. Recheck the alignment. Readjust and recheck the alignment until temperatures on each side of the guide lugs for both tracks feel about the same temperature.

Adjusting Track Tread Setting

JDPX1383

Fig. 8 — Narrow Track Tread Setting

JDPX1384

Fig. 9 — Wide Track Tread Setting

Tread setting is the measurement between the center of the tracks. It is recommended that the tread setting be set to match the crop spacing.

Configurations and setting procedures vary by manufacturer. One configuration uses an adjustable drive wheel hub and multiple track frame mounting positions to adjust the track width. Another configuration requires adding frame and axle spacers to achieve the desired track width.

Diagnosis of Wear

MID ROLLER AND FRONT IDLER PIVOT WEAR

The condition of mid rollers and the front idler pivot should be checked with the drive wheel on a raised surface such as a 4 x 4.

When the track is detensioned, the front idler pivot should allow the front idler to rock backward and allow the top part of the rubber track sag. No pivot movement of the front idler may indicate possible seized pivot pin/bushings

When the track is allowed to sag, the mid rollers should be inspected. An out-of-round mid-roller may cause vibration or uneven track operation/wear. Mid rollers should rotate freely.

TRACK WEAR

Inspect tread lugs and guide lugs for wear or cuts. If unusual chunking, chipping or cracking of lugs exist, the track may need to be repaired or replaced.

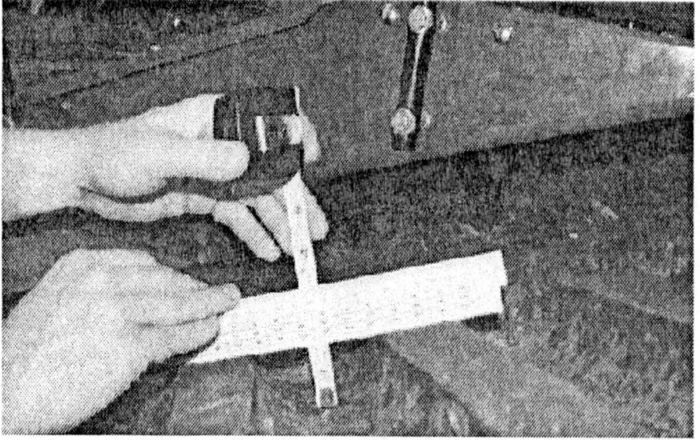

JDPX1387

Fig. 10 — Measurement of Track Wear

Tracks should be replaced when 20% of original tread lug remains. Measure as shown in Fig. 10. Replace track when lug measures less than 0.5 in. (13 mm).

Rubber Track Failures

RUBBER CHECKS, CRACKS AND CUTS

Smaller checks or cracks may develop on the track surfaces. These are usually caused by exposure to sunlight, ozone, air drafts, or electrical discharges. Normally this is an appearance condition only and will not affect the service life of the tracks. Protection from the elements will help prevent this condition. Even a coating of mud on the track may be helpful.

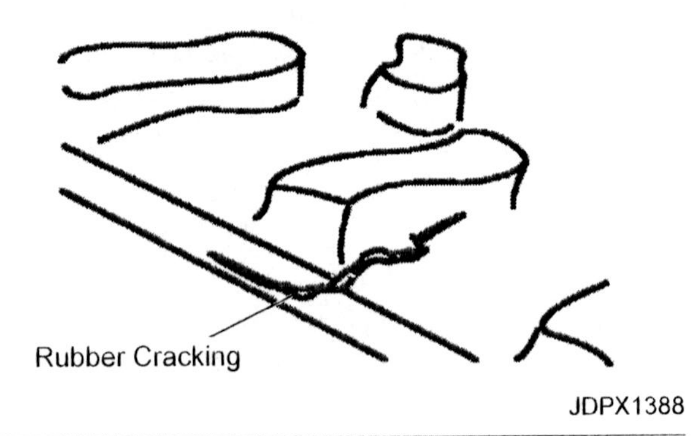

Fig. 11 — Tread Cracks

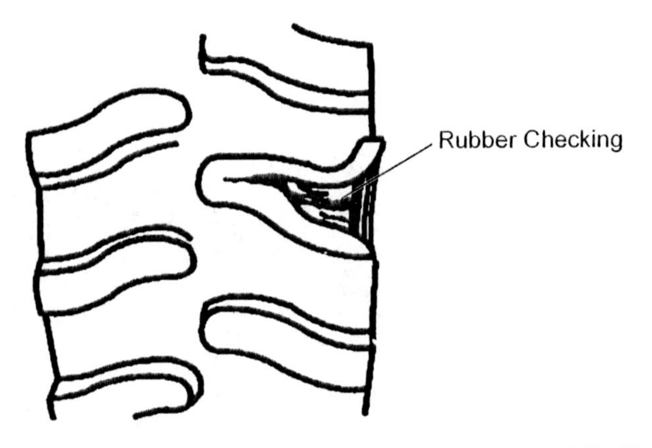

Fig. 12 — Large Check in Tread Lug

Larger cuts, checks and cracks in the tread may lead to deterioration and premature failure of the track. Small rocks and dirt may get into the tread, and if neglected, will eventually be pounded into the track plies. To prevent this, clean out the cut with an awl or similar tool to remove stones, etc. Then, cut away the rubber around the damage to form a cone shape extending to the bottom of the injury.

GREASE AND OIL DAMAGE

Keep grease, oil and petroleum products off tracks, as they destroy rubber. After using a machine for spraying, wash off any chemicals that may have dropped on the tracks.

UNEVEN WEAR FROM TRACK SLIP

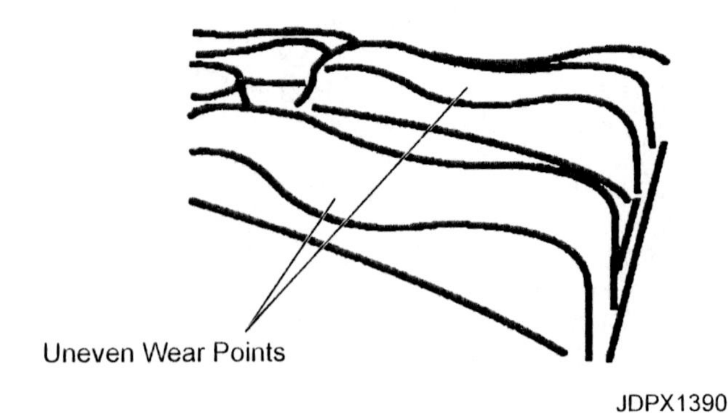

Fig. 13 — Uneven Tread Wear

Uneven wear occurs to some degree on all tracks and is most noticeable on wider tracks due to flexing at edges where track rollers do not press directly on the track (Fig. 13).

A tractor with too little ballast will wear the tread bars or will snag and cut the bars when subjected to severe service on abrasive surfaces. Sudden engagement of the throttle also causes this type of tread wear. Tread bars are cut and worn on the leading edge.

To prevent track slip wear, add ballast, reduce the draft load where possible, and engage the throttle slowly when starting.

STUBBLE WEAR

Stubble wear on the tread rubber is caused by operating a track over crop stubble. To avoid or limit stubble damage, adjust the track setting so neither track rides over the stubble, and make an effort to not spin the tracks excessively when they do contact stubble.

TRACK OBSTRUCTIONS

A bent frame, step assembly or attachment may rub on the tracks. Sharp corners can gouge the track. Check for these obstructions often.

HARD ROAD OPERATION

NOTE: Tracks will show increased wear, similar in nature to tire-vehicles, when operated on hard surface roads.

Operating on hard surface roads will wear tracks faster than any other factor. Rubber tracks are designed to operate in the field where soil can conform to the tread design, and where all portions of the tread make contact with the soil. If you plan to travel a long distance on hard surfaces significant and uneven wear to the tread can be expected.

During turns, much of the track slides sideways due to the large footprint of the track. On highly abrasive, hard surfaces, this action wipes the rubber off the tread bars and wears them down prematurely. Tracks that have been roaded substantially will show sharp clean leading lug edges and more wear in the area of track where mid rollers, drive wheel and idler wheel run (Fig. 13).

Rubber Track Repairs

REPAIRING TRACKS

Modern dealers and repair facilities are capable of repairing damaged and worn tracks. Repair of rubber tracks is possible except when extreme damage is present.

Rubber tracks can be relugged on the interior or exterior of the track. In some field operations, relugging pays big dividends. The deciding factor is the severity of the job the track must do. High speed and overload operations all reduce the life span of the track carcass.

If relugging or replacement of one track is necessary, it may be paired with a used track. The used track does not impair the pulling capability of the tractor. If the used track has less than half of its tread remaining, some loss of traction in loaded turns may become evident when powering a turn with the used track.

Rubber Track Removal and Installation

REMOVAL

To remove the rubber track:

1. Park tractor on a firm, level surface.

CAUTION: Avoid possible personal injury. Tractor must be fully supported and stable using support stands of sufficient capacity before working on tracks.

2. Raise entire tractor, keeping tractor level front to back. Lift tractor until approximately 7 in. (175 mm) clearance exists between belt tread and ground.

3. Install support stands approved for this purpose to front frame support and differential case.

4. Loosen but do NOT remove cap screws on outer idler wheel.

5. Detension the track.

NOTE: The track should be dragging on the ground.

6. Remove the outer idler wheel.

7. Remove the track.

INSTALLATION

1. Install the track onto drive wheel.

2. Slide front of track over inside idler wheel.

NOTE: Use of soapy water on the idler wheel greatly aids in getting the track over the wheel.

3. Install the outer idler wheel and hand-tighten cap screws.

4. Tension the track.

5. Tighten outer idler wheel cap screws.

6. Remove support stands and lower tractor to the ground.

7. Align the track.

8. Retighten the idler wheel cap screws to specification.

9. Tension the track a second time, to ensure full track tension.

10. Remove track tension hose.

Ballast

Modern rubber-track-equipped tractors are designed with adequate horsepower for their field operations. However, the machine weight by itself may not be sufficient for full traction and drawbar pull, resulting in track slippage. The effective weight on the track determines how much a machine can pull, depending on the ground surface.

The effective weight is the total weight of the machine (including added ballast) plus the weight that is transferred to the machine by an implement.

Since there is generally little control over the operating surface, slippage must be controlled by adding weight to the machine.

WHY BALLAST?

Tracks develop pull in relation to the amount of weight on them. The greater the weight, the more tractive effort the track can exert. However, weight compacts soil and increases the rolling resistance. The heavier the tractor, the more power it needs to propel itself across the soil surface. There is, therefore, an optimum weight for a given tractor under given conditions. The use of proper ballast is normally the best way to obtain this optimum weight.

BALLASTING FOR TRACTIVE PERFORMANCE

A guide in ballasting tractors is to use the kind of ballast needed for a particular job — light, medium or heavy ballast. Start the process of ballasting with the lightest ballast that may do the job. Then add ballast as needed to get the performance desired.

NOTE: Correct weight balance must be maintained when adding or removing ballast for best tractive performance.

Ballast levels are based on the following travel speeds:

- Light — 5.5 mph (8.8 km/h) or more
- Medium — 4.5-5.5 mph (7.2-8.8 km/h)
- Heavy — less than 4.5 mph (7.2 km/h)

More or less weight will be needed if different travel speeds are used. Higher speeds do not require as much weight. The final indication of correct ballast is slip measured in the field.

NOTE: Radar is recommended to continually monitor slip. Checking slip manually is possible but will only show slip in one area of the field. Field conditions vary and maintaining the correct average slip is necessary to obtain optimum performance.

Optimum tractive performance of 2-5 percent slip is recommended. Add more weight if slip is excessive. Remove weight if there is less than minimum percent slip.

If implement pull at full load is 5.5 mph (8.8 km/h) or more, the tractor may operate unballasted. Medium ballast is a better choice, if operating at full load between 4.5 and 5.5 mph. Heavy ballast should only be used for the few implements (such as deep rippers) which require full-load traction below 4.5 mph.

BALLAST LIMITATIONS

IMPORTANT: **To extend drive train life and avoid excessive soil compaction and rolling resistance, avoid adding too much ballast. Never add ballast that results in continuous full-power loads below 4.1 mph (6.6 km/h).**

IMPORTANT: **Do not overload tractor. If maximum recommended weight is not enough, reduce load.**

Upon adding or removing ballast, also adjust front ballast for the particular operating condition.

Checking Front-to-Rear Balance

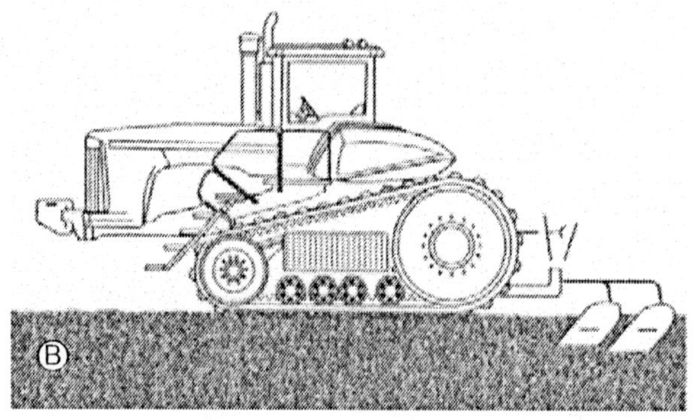

JDPX1391

Fig. 14 — Visual Front-to-Rear Balance Check

Correct front-to-rear balance should be maintained whenever ballast is added or removed. A good way to check for correct balance is driving or walking alongside of tractor during operation and visually checking to ensure entire contact area of track is in contact with soil surface at all times with implement engaged (B).

Adjustments may need to be made to the implement gauge wheels or more front weight may have to be added if daylight can be seen between the front idler wheel and the soil surface (A).

HOW MUCH TRACK SLIP?

Weight guides tell the amount of tractor weight required for a specific tractor under given conditions. However, they only give an estimate. The final criteria for adding ballast is the amount of travel reduction (% slip) of the drive wheels.

Under normal field conditions, travel reduction should be around 2-5 percent. Add more weight if travel reduction is greater than 5 percent. If there is less than 2 percent slip, weight should be removed. Zero slip is not good.

HOW TO MEASURE TRAVEL REDUCTION (TRACK SLIP)

For a tractor without a performance monitor, the travel reduction or wheel slip can easily be found by the following method:

NOTE: Track slip can be easily determined with the performance monitor or vehicle monitor, if equipped with optional radar unit.

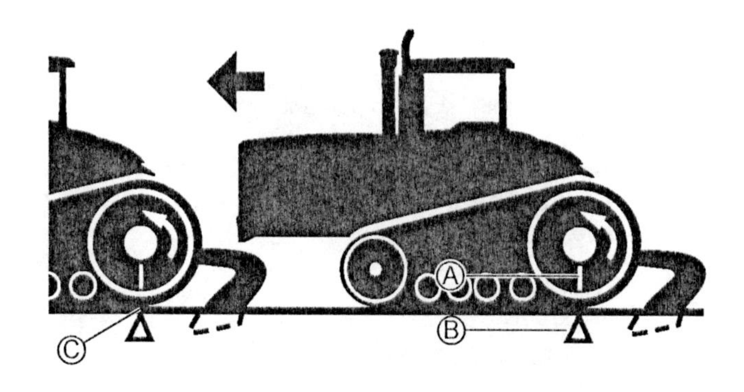

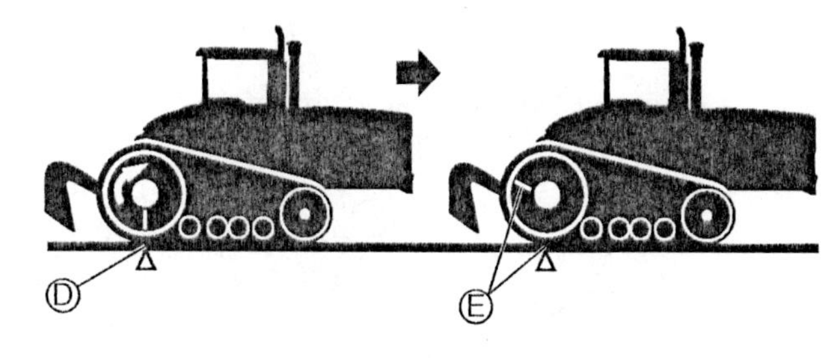

JDPX1392/JDPX1393

Fig. 15 — Measuring Travel Reduction Manually

1. Mark drive wheel (A).
2. With the tractor working, mark a starting point on the ground (B).

CAUTION: Avoid possible personal injury. Stay clear of moving tractor during this procedure.

3. Have an assistant follow tractor and mark ground where drive wheel completes 10 full revolutions (C).
4. At same working speed go back with implement raised. Line up the drive wheel mark with ground mark, or make a new mark on the drive wheel (D).

5. With tractor proceeding in a straight line with implement raised, count the revolutions of the drive wheel between the two marks on the ground to the nearest 1/4 track revolution (E).

6. Use the second count and the chart to determine slippage.

Revolutions	% Slip	Result
10	0	Remove Ballast
9.9	1	
9.75	2	Proper Ballast
9.5	5	
9.25	7	Add Ballast
9	10	

7. Adjust ballast to give 2 to 5 percent slippage.

If less than 9.25 revolutions is obtained, *weight should be added.* If more than 9.75 revolutions is obtained, *weight should be removed.*

NOTE: The formula for calculating track slip is as follows:

R^1 = *Revolutions with no load over a given distance*

R^2 = *Revolutions with load over the same distance*

$$\text{Percent Slip} = \frac{R^2 - R^1}{R^2}$$

Test Yourself

QUESTIONS

1. How do you check for proper alignment of a rubber track?

2. T/F — Rubber tracks must be replaced in pairs.

3. What factor causes faster wear than any other?

4. Describe the proper method to check front-to-rear balance on a rubber track tractor.

5. For best performance, what is the ideal amount of track slip?

   a. Zero slip.

   b. 2-5 percent.

   c. 5-10 percent.

6. What are two common configurations used to adjust track tread settings?

7. What are the main components of the tension mechanism?

8. T/F — Tensioning the track must be completed with tractor raised off the ground.

ANSWERS

1. Drive tractor in a straight line for 1300 ft (400 m). Feel for surface temperature difference on the sides of the guide lugs about 3/4 in. (20 mm) from the point where the lugs meet the track. A difference in temperature indicates the need to adjust alignment.

2. False — New tracks may be paired with a used track.

3. Operating on hard surfaces will wear tracks faster than any other factor.

4. Drive or walk alongside of tractor during operation. Visually check to ensure entire contact area of track is in contact with soil surface at all times with implement engaged.

5. B — 2-5 percent.

6. Configuration 1: Adjustable drive wheel hub and multiple track frame mounting positions.

   Configuration 2: Frame and axle spacers.

7. Track tension cylinder, accumulator, tension sensor, front idler wheel hub.

8. False — Tractor should be sitting on flat/level ground.

APPENDIX

Measurement Conversion Chart

Metric to English

LENGTH
1 millimeter = 0.03937 inches . . . . in
1 meter = 3.281 feet . . . . ft
1 kilometer = 0.621 miles . . . . mi

AREA
1 meter^2 = 10.76 feet^2 . . . . ft^2
1 hectare = 2.471 acres . . . . acre
(hectare = 10,000 m^2)

MASS (WEIGHT)
1 kilogram = 2.205 pounds . . . . lb
1 tonne (1000 kg) = 1.102 short ton . . . . sh tn

VOLUME
1 meter^3 = 35.31 foot^3 . . . . ft^3
1 meter^3 = 1.308 yard^3 . . . . yd^3
1 meter3 = 28.38 bushel . . . . bu
1 liter = 0.02838 bushel . . . . bu
1 liter = 1.057 quart . . . . qt

PRESSURE
1 kilopascal = 0.145 pound/in^2 . . . . psi

STRESS
1 megapascal or
1 newton/millimeter^2 = 145 pound/in^2 . . . . psi
(1N/mm^2 = 1MPa)

POWER
1 kilowatt = 1.341 horsepower (550 lb-ft/s) . . . . hp
(1 watt = 1 N•m/sec)

ENERGY (WORK)
1 joule = 0.0009478 British Thermal Unit . . . . BTU
(1 J = 1 W s)

FORCE
1 newton = 0.2248 pounds force . . . . lb force

TORQUE OR BENDING MOMENT
1 newton meter = 0.7376 foot-pound . . . . lb-ft

TEMPERATURE
$t_C = (t_F - 32)/1.8$

English To Metric

LENGTH
1 inch = 25.4 millimeters . . . . mm
1 foot = 0.3048 meters . . . . m
1 yard = 0.9144 meters . . . . m
1 mile = 1.608 kilometers . . . . km

AREA
1 foot^2 = 0.0929 meter^2 . . . . m^2
1 acre = 0.4047 hectare . . . . ha
(hectare = 10,000 m^2)

MASS (WEIGHT)
1 pound = 0.4535 kilograms . . . . kg
1 ton (2000 lb) = 0.9071 tonnes . . . . t

VOLUME
1 foot^3 = 0.02832 meter^3 . . . . m^3
1 yard^3 = 0.7646 meter^3 . . . . m^3
1 bushel = 0.03524 meter^3 . . . . m^3
1 bushel = 35.24 liter L
1 quart = 0.9464 liter L
1 gallon = 3.785 liter L

PRESSURE
1 pound/inch^2 = 6.895 kilopascals . . . . kPa
1 pound/inch^2 = 0.06895 bars . . . . bar

STRESS
1 pound/in^2 (psi) = 0.006895 megapascal . . . . MPa
or newton/mm^2 . . . . N/mm^2
(1 N/mm^2 = 1 MPa)

POWER
1 horsepower (550 lb-ft/s) = 0.7457 kilowatt . . . . kW
(1 watt = 1 N•m/s)

ENERGY (WORK)
1 British Thermal Unit = 1055 joules . . . . J
(1 J = 1 W s)

FORCE
1 pound = 4.448 newtons . . . . N

TORQUE OR BENDING MOMENT
1 pound-foot = 1.356 newton-meters . . . . N•m

TEMPERATURE
$t_F = 1.8 \times t_C + 32$

Metric Fastener Torque Values

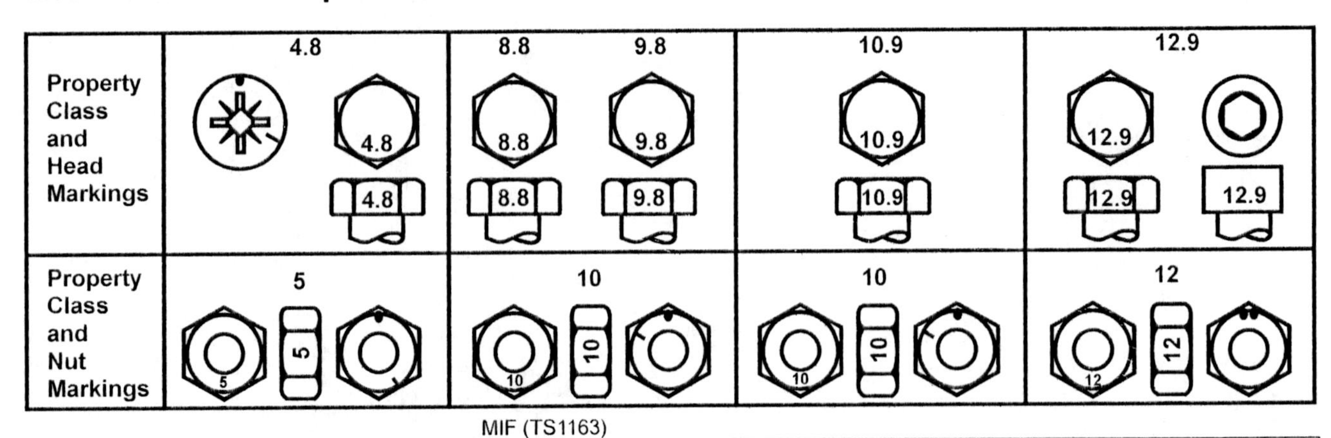

MIF (TS1163)

	Class 4.8				Class 8.8 or 9.8				Class 10.9				Class 12.9			
	Lubricated[a]		Dry[a]		Lubricated[a]		Dry[a]		Lubricated[a]		Dry[a]		Lubricated[a]		Dry[a]	
SIZE	N•m	lb-ft	N•m	lb-ft	N•m	lb-ft	N•m	lb-ft	N•m	lb-ft	N•m	lb-ft	N•m	lb-ft	N•m	lb-ft
M6	4.8	3.5	6	4.5	9	6.5	11	8.5	13	9.5	17	12	15	11.5	19	14.5
M8	12	8.5	15	11	22	16	28	20	32	24	40	30	37	28	47	35
M10	23	17	29	21	43	32	55	40	63	47	80	60	75	55	95	70
M12	40	29	50	37	75	55	95	70	110	80	140	105	130	95	165	120
M14	63	47	80	60	120	88	150	110	175	130	225	165	205	150	260	109
M16	100	73	125	92	190	140	240	175	275	200	350	225	320	240	400	300
M18	135	100	175	125	260	195	330	250	375	275	475	350	440	325	560	410
M20	190	140	240	180	375	275	475	350	530	400	675	500	625	460	800	580
M22	260	190	330	250	510	375	650	475	725	540	925	675	850	625	1075	800
M24	330	250	425	310	650	475	825	600	925	675	1150	850	1075	800	1350	1000
M27	490	360	625	450	950	700	1200	875	1350	1000	1700	1250	1600	1150	2000	1500
M30	675	490	850	625	1300	950	1650	1200	1850	1350	2300	1700	2150	1600	2700	2000
M33	900	675	1150	850	1750	1300	2200	1650	2500	1850	3150	2350	2900	2150	3700	2750
M36	1150	850	1450	1075	2250	1650	2850	2100	3200	2350	4050	3000	3750	2750	4750	3500

a. *"Lubricated" means coated with a lubricant such as engine oil, or fasteners with phosphate and oil coatings. "Dry" means plain or zinc plated (yellow dichromate - Specification JDS117) without any lubrication.*

DO NOT use these hand torque values if a different torque value or tightening procedure is given for a specific application. Torque values listed are for general use only and include a ±10% variance factor. Check tightness of fasteners periodically. DO NOT use air powered wrenches.

Shear bolts are designed to fail under predetermined loads. Always replace shear bolts with identical grade.

Fasteners should be replaced with the same class. Make sure fastener threads are clean and that you properly start thread engagement. This will prevent them from failing when tightening.

When bolt and nut combination fasteners are used, torque values should be applied to the NUT instead of the bolt head.

Tighten toothed or serrated-type lock nuts to the full torque value.

Reference: JDS-G200.

Metric Fastener Torque Values - Grade 7

	Steel or Gray Iron Torque		Aluminum Torque	
SIZE	N•m	lb-ft	N•m	lb-ft
M6	11	8	8	6
M8	24	18	19	14
M10	52	38	41	30
M12	88	65	70	52
M14	138	102	111	82
M16	224	165	179	132

Inch Fastener Torque Values

SAE Grade and Head Markings	No Marks — 1 or 2[a]	5 / 5.1 / 5.2	8 / 8.2
SAE Grade and Nut Markings	No Marks — 2	5	8

MIF (TS1162)

	Grade 1				Grade 2[a]				Grade 5, 5.1 or 5.2				Grade 8 or 8.2			
	Lubricated[b]		Dry[b]		Lubricated[b]		Dry[b]		Lubricated[b]		Dry[b]		Lubricated[b]		Dry[b]	
SIZE	N•m	lb-ft	N•m	lb-ft	N•m	lb-ft	N•m	lb-ft	N•m	lb-ft	N•m	lb-ft	N•m	lb-ft	N•m	lb-ft
1/4	3.7	2.8	4.7	3.5	6	4.5	7.5	5.5	9.5	7	12	9	13.5	10	17	12.5
5/16	7.7	5.5	10	7	12	9	15	11	20	15	25	18	28	21	35	26
3/8	14	10	17	13	22	16	27	20	35	26	44	33	50	36	63	46
7/16	22	16	28	20	35	26	44	32	55	41	70	52	80	58	100	75
1/2	33	25	42	31	53	39	67	50	85	63	110	80	120	90	150	115
9/16	48	36	60	45	75	56	95	70	125	90	155	115	175	130	225	160
5/8	67	50	85	62	105	78	135	100	170	125	215	160	215	160	300	225
3/4	120	87	150	110	190	140	240	175	300	225	375	280	425	310	550	400
7/8	190	140	240	175	190	140	240	175	490	360	625	450	700	500	875	650
1	290	210	360	270	290	210	360	270	725	540	925	675	1050	750	1300	975
1-1/8	470	300	510	375	470	300	510	375	900	675	1150	850	1450	1075	1850	1350
1-1/4	570	425	725	530	570	425	725	530	1300	950	1650	1200	2050	1500	2600	1950
1-3/8	750	550	950	700	750	550	950	700	1700	1250	2150	1550	2700	2000	3400	2550
1-1/2	1000	725	1250	925	990	725	1250	930	2250	1650	2850	2100	3600	2650	4550	3350

a. *"Grade 2" applies for hex cap screws (not hex bolts) up to 152 mm (6-in.) long. "Grade 1" applies for hex cap screws over 152 mm (6-in.) long, and for all other types of bolts and screws of any length.*

b. *"Lubricated" means coated with a lubricant such as engine oil, or fasteners with phosphate and oil coatings. "Dry" means plain or zinc plated (yellow dichromate - Specification JDS117) without any lubrication.*

DO NOT use these hand torque values if a different torque value or tightening procedure is given for a specific application. Torque values listed are for general use only and include a ±10% variance factor. Check tightness of fasteners periodically. DO NOT use air powered wrenches.

Shear bolts are designed to fail under predetermined loads. Always replace shear bolts with identical grade.

Fasteners should be replaced with the same grade. Make sure fastener threads are clean and that you properly start thread engagement. This will prevent them from failing when tightening.

When bolt and nut combination fasteners are used, torque values should be applied to the NUT instead of the bolt head.

Tighten toothed or serrated-type lock nuts to the full torque value.

Reference: JDS-G200.

INDEX

U

V

W

Z